From Proteins to Proteomics

Proteomics aims to study all the proteins of human and other living systems, and their properties to provide an integrated view of cellular processes. The proteomic study involves the application of rapidly evolving high-throughput technologies and new platforms that are coming forward regularly, providing versatile novel tools for biomedical and pharmaceutical applications.

This book provides a detailed understanding of the basics of proteins and proteomics, gel-based proteomics techniques, basics of mass spectrometry and quantitative proteomics, interactomics: basics and applications, and advancements in proteomics. It also covers basic knowledge about sample preparation, mass spectrometry workflow, different chromatography technologies, and quantitative proteomics.

The text highlights the application and challenges of various high-throughput integrated proteomics technologies capable of fast and accurate screening of thousands of biomolecules that are found to be very effective for studying disease pathobiology and identification of next-generation biomarkers and potential drug/vaccine targets; and are therefore considered valuable tools for multidisciplinary research.

Features

- Basics of proteins and proteomics techniques
- In-depth understanding of mass spectrometry and quantitative proteomics
- An overview of interactomics and its application for translational research
- Advancements in the field of proteomics and challenges in clinical applications

We hope the knowledge gained from reading this book will intrigue and motivate young minds to explore future opportunities in the constantly evolving field of proteomics.

From Proteins to Proteomics
Basic Concepts, Techniques, and Applications

Sanjeeva Srivastava

Professor
Department of Biosciences and Bioengineering
Indian Institute of Technology
Bombay, India

CRC Press
Taylor & Francis Group
Boca Raton London New York

CRC Press is an imprint of the
Taylor & Francis Group, an **informa** business

First edition published 2023
by CRC Press
6000 Broken Sound Parkway NW, Suite 300, Boca Raton, FL 33487-2742

and by CRC Press
2 Park Square, Milton Park, Abingdon, Oxon, OX14 4RN

© 2023 Taylor & Francis Group, LLC

CRC Press is an imprint of Taylor & Francis Group, LLC

Library of Congress Cataloging-in-Publication Data

Names: Srivastava, Sanjeeva, author.
Title: From proteins to proteomics : basic concepts, techniques, and applications / authored by Sanjeeva Srivastava.
Description: First edition. | Boca Raton : CRC Press, 2023. | Includes bibliographical references and index.
Identifiers: LCCN 2022013971 (print) | LCCN 2022013972 (ebook) | ISBN 9780367566173 (hardback) | ISBN 9780367566203 (paperback) | ISBN 9781003098645 (ebook)
Subjects: MESH: Proteomics--methods.
Classification: LCC QP551 (print) | LCC QP551 (ebook) | NLM QU 460 | DDC 572/.6--dc23/eng/20220727
LC record available at https://lccn.loc.gov/2022013971
LC ebook record available at https://lccn.loc.gov/2022013972

ISBN: 978-0-367-56617-3 (hbk)
ISBN: 978-0-367-56620-3 (pbk)
ISBN: 978-1-003-09864-5 (ebk)

DOI: 10.1201/9781003098645

Typeset in Times
by KnowledgeWorks Global Ltd.

Contents

Module V   Advancements in Proteomics

Preface

This book introduces the basic biology of proteins and the advancing science called proteomics which looks into the protein properties from a global perspective. Proteomics aims to study all the proteins of human and other living systems and their properties to provide an integrated view of cellular processes. Proteomics study involves the application of high-throughput technologies that are rapidly evolving and new platforms are coming forward regularly with versatile novel applications and are valuable tools for biomedical and pharmaceutical applications. This book will be beneficial for undergraduate and postgraduate students having a basic knowledge of biology and will provide them with a detailed understanding of proteins, proteomics technology, and its applications.

The concept and content of this book arose from the National Programme on Technology Enhanced Learning (NPTEL) online course on "Introduction to Proteomics" conducted by Prof. Sanjeeva Srivastava and initiated by the Indian Institute of Technology Bombay. The NPTEL has offered open online courses (https://onlinecourses.nptel.ac.in/noc20_bt20/preview) since 2014 along with certificates from the IITs/ IISc for those that complete the courses successfully. This course is an extensive 8-week program with around 2000 participants every year including graduate students, researchers, as well as faculty. In addition, the YouTube channel for NPTEL is the most subscribed educational channel, with 1 billion views and 31+ lakh subscribers. The enormous enthusiasm from the participants led to this book material which is easily accessible to all the students and researchers that are beginners in this field.

This book is divided into five modules providing a detailed understanding of the basics of proteins and proteomics, gel-based proteomics techniques, basics of mass spectrometry and quantitative proteomics, interactomics: basics and applications, and advancements in proteomics. Module I includes two chapters describing the basics of amino acids, the backbone of the protein molecule, and also covers the proteomics evolution that emerged from protein chemistry solely due to technological advancements.

Modules II and III cover in detail the two major aspects of proteomics, gel-based proteomics, and mass spectrometry-based proteomics. The gel-based module presents different techniques like SDS-PAGE, 2-DE, 2D-DIGE, and so on. These techniques have had a major contribution to the transition from protein chemistry to proteomics. Mass spectrometry, on the other hand, is an advanced analytical technique for accurate mass measurement. These chapters also discuss the basics of mass spectrometry, sample preparation, liquid chromatography, hybrid mass spectrometers, and quantitative proteomics techniques such as iTRAQ, SILAC, and TMT using mass spectrometry. The course also provides basic knowledge about sample preparation, mass spectrometry workflow, different chromatography technologies, and quantitative proteomics.

Module IV focuses on interactomes, providing a better understanding of genomes and proteomes, which govern the network of interactions in a cell. This module also discusses various technology platforms such as protein microarrays, label-free biosensors, and surface plasmon resonance. Module V highlights the application and challenges of various high-throughput integrated proteomics technologies capable of fast and accurate screening of thousands of biomolecules that are found to be very effective for studying disease pathobiology and identification of next-generation biomarkers and potential drug/ vaccine targets, and therefore considered valuable tools for multidisciplinary research.

This book thus aims to provide readers with the basic knowledge of proteins, proteomics, and translation of proteomics knowledge for clinical application. We hope the knowledge gained from the reading of the text will intrigue young minds to explore future opportunities in the constantly evolving field of proteomics.

Author

Sanjeeva Srivastava is a Professor and the Group Leader of the Proteomics Laboratory at the Department of Biosciences and Bioengineering at the Indian Institute of Technology, Bombay. High-throughput proteomics, protein microarrays, and mass spectrometry are among his specialties. He has implemented groundbreaking AI-driven data analytics on big biological datasets. He has been at the forefront of biomedical research based on big data. His group's current research focuses on the development of clinical biomarkers for infectious diseases and malignancies. His group also pioneered therapeutic target identification efforts and decoded protein interaction networks in human illnesses such as gliomas and contagious diseases including COVID-19 and malaria. His group has developed reliable diagnostic biomarkers and described the pathophysiology of severe malaria (falciparum and vivax) and COVID-19, especially the underlying mechanisms that lead to the development of severe sequelae. Dr. Srivastava is an active contributor to global proteomics research and development. He serves on the Executive Committee of Human Proteome Organization (HUPO) and Proteomics Society, India (PSI). He has more than 150 publications from his work as an independent researcher at IIT Bombay. To date, he has filed 15 patents that include biomarkers for various types of cancers; infectious diseases like malaria, leptospirosis, COVID-19; and methods for improvement in uncharted territories such as fish and plant proteomics.

Module I

Basics of Proteins and Proteomics

1

Basics of Amino Acids and Proteins

Preamble

Proteins are the most complex and versatile macromolecules comprising amino acids as the building blocks. There are 20 standard amino acids, with the number varying according to the side chain or functional group connected to the amino acid's central asymmetric carbon atom. Each amino acid is encoded by a codon (which is a three-nucleotide stretch) through the process of translation of the corresponding mRNA transcript, which in turn is generated from the DNA sequence by transcription. Multiple amino acids are joined together in different combinations by means of peptide bond between two amino acids, to form a peptide. Multiple peptides are joined together to form polypeptides, which are eventually joined to form protein. Usually, many proteins are composed of different subunit complexes, which may be structurally and functionally different. The hierarchical arrangement can be depicted as:

DNA—mRNA—Codons—Amino acids—Peptides—Polypeptides—Proteins

To understand the protein structure and function, it is important to understand the basics of amino acids, their structure and function, their properties with respect to their microenvironment. Knowledge of these fundamental aspects can be further employed to gain deeper insights using proteomic approaches.

Terminology

- **Enantiomers:** The *dextro* and *levo* rotatory forms of given amino acids, which are nonsuperimposable mirror images of each other.
- **Amino group:** This comprises of a covalently linked NH_2 group to the primary carbon atom. It occurs as either NH_2 or NH_3^+ depending on the pH of the surrounding medium. All other amino acids only contain primary amino groups, with the exception of proline, which possesses a secondary amino group.
- **Carboxyl group:** The central alpha (α) carbon atom is covalently bonded to a COOH group, which, depending on the pH of the medium, can either be COOH or COO^-.
- **Peptide bond:** It is also referred to as the amide bond, and it holds two amino acids together. In a peptide bond, a covalent bond—the creation of which requires the loss of a water molecule—binds the carboxyl group of one amino acid to the amino group of the other amino acid.
- **Psi(ψ) and Phi(φ) angles:** Angles between the α-Carbon and Carbon of carboxyl group and, between the α-Carbon and Nitrogen of amino group, respectively. The advantageous protein conformations will be decided by these angles.
- **Hydrogen bonds:** Interaction between hydrogen atom and any other electronegative atom when placed in closed vicinity to each other is known as Hydrogen bond. It is just a weak attractive force and may develop within or between polypeptide chains.
- **Electrostatic interactions:** Attractive interactions between groups of atoms with opposing charges that help maintain the structure of proteins.
- **Hydrophobic interactions:** Non-specific interactions between non-polar amino acid side chains, which act to bury these hydrophobic residues away from a polar environment.
- **Van der Waals forces:** Weak attractive or repellent forces generated on by polarization shifts.
- **Disulfide bridges:** Different cysteine residues in the polypeptide chain(s) interact specifically and undergo oxidation process, forming disulphide bonds.

DOI: 10.1201/9781003098645-2

1.1 Basics of Amino Acids

1.1.1 Structural Backbone of Amino Acids and Their Classification

Amino acids are the building blocks or monomers that make up the proteins. An amino acid except glycine is composed of an asymmetric or chiral central carbon atom, which is bound to four different groups, namely: a carboxyl group (—COO), an amino group (—NH3+), a hydrogen atom (—H), and a side chain (—R). The central carbon atom is known as the α-carbon atom. The amino acids with a side chain bonded to this carbon are simply referred to as α-amino acids. These are basic monomeric constituents of proteins found in varying amounts depending upon the type of protein. They are classified based on the properties of their side chains or R groups that vary in size, structure, and charge. The polarity of the side chains is one of the main basis for this classification.

Essential amino acids are those that cannot be synthesized *de novo* in an organism and therefore must be included in the diet. On the other hand, non-essential amino acids may be synthesized from a variety of precursors. The amino acids glycine, alanine, proline, valine, leucine, isoleucine, and methionine all have non-polar aliphatic side chains. Polar but uncharged side chains make up the amino acids serine, threonine, asparagine, glutamine, and cysteine. Positively charged side chains can be found on lysine, arginine, and histidine. The amino acids glutamic acid and aspartic acid are polar and negatively charged. There are three essential amino acids with aromatic side chains: tryptophan, tyrosine, and phenylalanine.

1.1.2 Isomerism in Amino Acids

The term "chirality" arises from the Greek term "cheir" meaning "handedness". Just like the two hands are nonsuperimposable mirror images of each other, amino acid molecules are also nonsuperimposable due to their chiral α-carbon center. Except for glycine, all amino acids can exist as L-isomers or D-isomers as they contain an asymmetric center that makes them chiral in nature due to which they can rotate the plane of polarized light. The two enantiomers are designated as dextro and levo, which rotate the plane of polarization in opposite directions. The two enantiomers of amino acids are nonsuperimposable mirror images due to the spatial arrangement of four different groups around the chiral carbon atom. Rotation of either isomer about its central axis will not give rise to the other isomeric structure. However, with respect to conformational stability, all amino acids are usually found in the L-isomeric form in humans.

1.1.3 Acidic and Basic Properties of Amino Acids

Amino acids are amphoteric in nature, i.e., they act both as acidic and basic due to the presence of two amine and a carboxylic group. Hence, they make a perfect example of zwitterions. A zwitterion is a molecule with functional groups with at least one positive and one negative electrical charge so that the net charge is zero. Amino acids vary in their acid-base properties based on the functional group (R) attached to them and have characteristic titration curves. Amino acids in an acidic medium exist in the completely protonated form carrying a net positive charge, which can be confirmed by simple paper electrophoresis. The sample solution is applied at the center of the strip and current is passed through it. The colorless amino acid solution can be detected by spraying the strip with ninhydrin which gives it a purple color. Migration of the spot toward the negatively charged cathode confirms the net positive charge of the amino acid.

All amino acids exhibit a characteristic titration curve with distinct pKa values. 0.1 N NaOH is added to the acidic amino acid solution. The cationic form of the amino acid is gradually converted into its neutral or zwitterionic form by the loss of a proton from its COOH group. This can again be confirmed by electrophoresis where there is no migration of the sample spot. The number of equivalents of alkali being consumed is plotted against the pH of the amino acid solution to obtain the titration curve. pKa1 of glycine is found to be 2.34, i.e., it starts to lose its carboxyl group proton at this pH. The second stage of the titration curve is the proton's removal from the amino group. Continued addition of alkali to the amino acid solution gradually converts the zwitterionic form into anionic form which can be confirmed by the migration of the sample spot toward the anode during electrophoresis. The pKa2 of an amino acid is obtained by continued addition of alkali to the neutral solution of the amino acid. pKa2 of glycine is found to be 9.6. Some amino acids having positively or negatively charged side chains will have pKa1, pKa2, and pKaR, which correspond to ionization of the side chain. These amino acids have a good buffering capacity of around 1 pH unit on either side of their pKa values.

1.1.4 Significance of Amino Acid Sequence

The significance of amino acid sequence was elucidated by Anfinsen's experiment (Figure 1.1), which is described as follows.

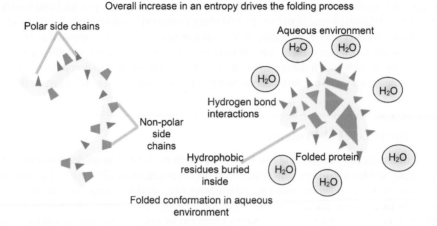

FIGURE 1.1   Amino acid structure determines 3D folding.

Anfinsen's experiment (Reported in 1973)

- **Objective:** To establish that amino acid sequence dictates protein folding.
- **Experimental insights:** The role of compounds like urea, guanidine HCl, beta (β)-mercaptoethanol, etc., on the refolding of an enzyme ribonuclease A was elucidated through the experimental study. 8M urea and guanidine HCl together act as denaturants and disrupt the non-covalent bonds present in the protein. β-Mercaptoethanol, on the other hand, acts as a reducing agent and disrupts the disulfide bonds, and when applied in large excess converts the disulfides to a sulfhydryl. This denatured and reduced form of ribonuclease A lacks enzymatic activity. When urea and β-mercaptoethanol are removed by dialysis, the denatured and reduced ribonuclease A spontaneously regains its native conformation and enzymatic activity. The removal of β-mercaptoethanol alone, however, did not render the ribonuclease A enzymatic activity. It was also observed that trace amounts of β-mercaptoethanol facilitated the correct disulfide bonding and thereby enhanced the formation of native ribonuclease A.
- **Conclusions:** The phenomenon of protein folding is governed by the amino acid sequence, which is the actual repository of all the genetic information.

Anfinsen's experiment. Ribonuclease A in its native state has four disulfide bonds among its cysteine residues. When treated with β-mercaptoethanol and 8M urea, the protein undergoes denaturation and the disulfide linkages are broken. Enzyme activity is lost in the denatured state. It was observed by Anfinsen that removal of urea and β-mercaptoethanol led to the refolding of the enzyme to resume its native state with more than 90% of enzyme activity being intact. However, if only β-mercaptoethanol was removed in presence of urea, the formation of disulfide bonds was random, resulting in an enzyme with only around 1% activity.

1.2 Peptide Bond

Amino acids are linked together such that the carboxyl group of one amino acid combines with the amino group of another using a condensation reaction with the loss of a water molecule to form a dipeptide. Many such amino acids are linked together forming a polypeptide. The peptide bond is rigid due to its partial double-bond character arising from resonance structures. However, the bonds between the α-carbon amino and carboxyl groups are purely single bonds that are free to rotate.

The covalent bond that holds two adjacent amino acid residues together is known as a peptide bond, which is formed between the carboxyl group of one amino acid and the amino group of other amino acids and is accompanied by the release of a water molecule. The peptide bond is stabilized by the resonance structure. The amide bond exhibits a partial double-bond character and is planar. In other words, it can exist in "cis" and "trans" forms. In the unfolded form of a given protein, the peptide bond has the liberty to take up either of the two forms; however, the folded conformation has the peptide bond in a single form only. The "trans" form is usually preferred as its conformation is stable as compared to the "cis" form (exception: proline, which can exist in "cis" as well as "trans" form). The psi and phi are the angles of rotation about the bond between the α-carbon atom and carboxyl and amino group, respectively. These angles determine which protein conformations will be favorable during protein folding (Baker, 2000).

1.3 Structural Level of Proteins

A peculiar arrangement of amino acid residues of a protein is essential to render it functionally active and structurally stable. A given protein structure is the organization of different L-α-amino acids in a particular way, which is aided by numerous interactions like covalent interactions, hydrogen bonding, van der Waals forces, ionic interactions and other hydrophobic interactions. The protein molecule can also undergo reversible changes in its structure to perform its function effectively, and these structural variations are termed as "conformational changes". Protein structure has four different levels (Table 1.1): primary, secondary, tertiary and quaternary.

1.3.1 Primary Structure

The primary structure of a protein constitutes amino acids that are arranged in a linear fashion to make a polypeptide chain. Amino acids are joined together in a head-to-tail arrangement through peptide bonds. The exact order of the amino acids in a specific protein is the primary sequence for that protein.

1.3.2 Secondary Structure

The secondary structure of proteins is mainly a function of the pattern of hydrogen bonds in the side chain groups and their characteristic geometry can be deciphered from the phi and psi angles. α-helices and β-sheets are the two most common forms in which proteins can exist in their secondary structure. The α-helices can be further classified as left-handed or right-handed depending on their orientation. However, most α-helices are right-handed since this conformation is energetically more favorable. β-sheets can be parallel or anti-parallel. Amino acids in parallel β-sheets, which run in the same direction, interact with two different amino acids on the adjacent strand through hydrogen bonds. Amino acids in anti-parallel strands, on the other hand, interact with only one amino acid on an adjacent strand.

A compact globular structure is required for all proteins to attain structural stability. This can be made possible only if there are turns or loops between the different secondary structures. β-turns, which are the most commonly observed turn structures, consist of rigid, well-defined structures that usually lie on the surface of the protein molecule and interact with other molecules. A combination of secondary structures such as the helix-turn-helix, which has two α-helices separated by a turn, is known as super-secondary structures or motifs.

Ramachandran plot can be used to understand the relative distribution of α-helices and β-sheets within a given protein depending on the phi and psi angles of the component amino acid residues. The range of phi and psi values can be mapped to various areas of the plot in order to discern distinct secondary structures. The permitted zones are those that depict the conformations in which polypeptide atoms would not clash sterically. Disallowed regions generally involve steric hindrance between side-chain and the main-chain atoms. Glycine stands out as an exception and has no side chain and therefore can adopt phi and psi angles in all four quadrants of the Ramachandran plot.

The α-helix has intra-chain hydrogen bonds between the "H" of NH and "O" of CO in every 4th residue. The amino acid proline that possesses a cyclic side chain does not fit into the regular α-helix structure and thereby limits the flexibility of the backbone. It is commonly referred to as the "helix breaker". α-helices can also wind around each other to form stable structures such that their hydrophobic residues are buried inside, while their polar side chains are exposed to the aqueous environment. The main protein found in hair, α-Keratin, is made up of two of these left-handed superhelix-shaped coiled coils. Collagen, which is a fibrous component as in skin and muscle, consists of three such coiled α-helices. Glycine appears every third residue in the distinctively recurrent amino acid sequence glycine-proline-hydroxyproline.

The β-pleated sheet discovered by Pauling and Corey is another common secondary structure with periodic repeating units. It is made up of two or more polypeptide chains with side chains that face upward and downward in relation to the plane. Unlike the α-helix, the β-sheet is formed by hydrogen bonds between polypeptide strands, rather than within a strand. Amino acids in parallel β-sheets, which run in the same direction, interact with two different amino acids on the adjacent strand through hydrogen bonds. Amino acids in anti-parallel strands, on the other hand, interact with only one amino acid on an adjacent strand. The amino acids in β-sheets are more extended than in α helix, with 3.5 Å between each adjacent residue. When two or more β-sheets are layered close together in a protein, the R group is relatively small to maintain stability. For instance, the spider web protein fibroin has a high content of glycine and alanine.

Almost all proteins exhibit a compact, globular structure which is possible only if there are turns or loops between the various regions. β-Turns, which are the most commonly observed turn structures, consist of rigid, well-defined structures that usually lie on the surface of the protein molecule and interact with other molecules.

Amino acids located far apart on the polypeptide chain interact with each other through hydrogen bonding, electrostatic interactions and disulfide bridges, etc., allowing the protein to fold three dimensionally in space, giving rise to the tertiary structure. Folding takes place such that the hydrophobic residues are buried inside the structure while the polar residues remain in contact with the surroundings.

1.3.3 Tertiary Structure

The interactions between amino acid side chains that are dispersed throughout the polypeptide sequence and produce a three-dimensional (3D) arrangement of amino acids characterize the tertiary structure of a protein. John Kendrew's analysis of the tertiary structure of myoglobin made it clearly evident that the position of amino acid side chains within the structure depends on their type. Hydrophobic residues are found buried inside the structure while the polar amino acids are found on the surface. Seventy percent of the main chain of myoglobin is folded into α-helices with the rest being turns and loops, which are essential to give it a compact structure.

1.3.4 Quaternary Structure

Many proteins have more than one polypeptide chain, also called a "subunit", that are assembled by various interactions like electrostatic, van der Waals, disulfide bonds, giving rise to the quaternary structure. Different polypeptide chains or subunits interact with one another and are kept together by van der Waals, ionic, and electrostatic interactions. The final level of protein structure, the quaternary structure, is claimed to be present in these multi-subunit proteins (Shea & Brooks, 2001).

The hemoglobin protein has a quaternary structure. It is composed of two α and two β-chains. Each gene in the α-globin gene locus on chromosome 16 participates in the production of the α-globin chain. All of the genes that are expressed from the time that hemoglobin is developing in the embryo through the time when the adult β-globin gene is expressed are found in the β-globin gene locus, which is located on chromosome 11. Ribosomes in the cytosol produce the globin chains. Each hemoglobin subunit is joined to a heme prosthetic group. Protoporphyrin, a complex organic ring structure, surrounds the central iron atom, which is present in its ferrous form. The hemoglobin's ability to bind oxygen depends on the heme group. Six coordination bonds are formed by the iron atom, with four of them being to the nitrogen atoms of porphyrin rings, one being to a His side chain in the globin subunit, and the other serving as the oxygen binding site. Iron does not bind to oxygen while it is in the Fe^{3+} form.

TABLE 1.1

An Overview of the Different Structural Levels of Proteins

Primary Structure	Secondary Structure	Tertiary Structure	Quaternary Structure
A stretch of amino acids joined together by amide or peptide bonds to form a linear polymer is known as primary structure of the protein. The amino acid sequence can be determined by techniques like Edman Degradation or Mass Spectrometry. The first primary structure that was deduced by Fredrick Sanger was that of insulin.	The secondary structure of proteins is defined by folding of a polypeptide backbone with the help of hydrogen bonds between proximal amino acids giving rise to a regular arrangement. *α-helices* and *β-sheets* are the most commonly observed secondary structures of proteins. The amino acid proline is known as the "helix breaker" as it tends to disrupt the helix and is often found at a bend in the structure known as reverse turns or bends.	The tertiary structure of protein is defined by interactions between amino acid side chains that are located far apart in the polypeptide sequence and result into formation of a 3D arrangement of the amino acids. The hydrophobic residues compose the core of the protein, whereas, the hydrophilic ones interact with the polar surroundings. The interactions that hold this 3D structure include hydrophobic bonds, electrostatic interactions, hydrogen bonds, etc.	Many proteins have more than one polypeptide chain, also called as subunit, that are assembled together by various interactions like electrostatic, van der Waals, disulfide bonds, etc., giving rise to the quaternary structure. The individual subunits of a quaternary structure when present in a peculiar conformation render the protein functional.

1.4 Protein Folding and Misfolding

Anfinsen's experiment in 1973 elucidated that amino acid sequence dictates the 3D structure of a protein. Protein folding results in tertiary structure and ensures that hydrophobic residues are buried inside. A protein when folded completely and has achieved its native conformation allows it to carry out its intricate biological functions. Improper protein aggregation and misfolding are corrected by the molecular chaperones (heat shock proteins in eukaryotes), which facilitate the process of folding in case of large proteins. The protein transitions through a higher free-energy state as it goes from an unfolded (high-energy state) to a native and folded (low-energy state) form. Protein folding is driven by a reduction in the number of hydrophobic side chains that are exposed to the surrounding aqueous environment, which is followed by a concurrent rise in the entropy of the constituent parts. The protein has a certain shape in its natural state, but when it is unfolded, it can adopt any of its several conformations. The folded proteins are further stabilized by intramolecular hydrogen bonding, the strength of which is highly influenced by the external environment. Protein folding is considered to be significant not only for generating functionally active proteins but also for other functions like cellular trafficking and regulation of cell cycle and cell growth.

The dire consequences of protein misfolding can turn out to be harmful to the cells and the organism in question (Table 1.2). Protein misfolding mainly occurs because of mutations or inappropriate covalent modifications that can lead to the formation of a 3D structure different from the actual native conformation of the protein (Vendruscolo et al., 2003). The misfolded protein may have altered functions or, more often than not, end up having harmful side effects. The accumulation of these misfolded proteins or peptide fragments leads to degenerative diseases, characterized by the formation of insoluble protein plaques in the liver or brain.

Thermodynamics of protein folding: An unfolded polypeptide chain has very high free energy and entropy. The formation of favorable connections and the assumption of a more stable state during protein folding have the effect of lowering the system's free energy. The entropy reduces as the protein continues to fold into its stable, low-energy native state shape. While initially appearing to be unfavorable for the system, it is important to keep in mind that the entropy of the water molecules in the area grows as a result of the process, raising the system's total entropy and making it more favorable and spontaneous.

The distribution of polar and non-polar amino acid residues within the protein controls how the protein folds. "Hydrophobic collapse" is the process through which hydrophobic amino acids are forced to interact with one another. As they come together, the water molecules in their immediate vicinity are eliminated. The hydrophobic residues are hidden inside the protein's core, while the polar residues stay on the surface and form hydrogen bonds with water molecules. Proteins often adopt just one distinct native state conformation, which is the least unstable and has the lowest free energy. Due to the side chain characteristics of amino acids, such as hydrophobicity, size, shape, etc., folding is restricted to a single conformation. In the extremely cooperative process of folding, the intermediates are steadily stabilized. Theoretically, it is feasible to infer protein structure from the amino acid sequence, but this is frequently constrained by a number of long-range interactions.

Molecular chaperones for protein folding: The unfolded protein is bound by DnaJ and then by DnaK, which is an adenosine triphosphate (ATP) bound protein. The hydrolysis of ATP into adenosine diphosphate (ADP) and inorganic phosphate (Pi) by DnaK is stimulated by DnaJ. The unfolded protein is still strongly linked to the resultant DnaK-ADP. ADP and DnaJ are made more readily available for release by bacteria due to the nucleotide exchange factor GrpE. DnaK is still attached to the partly folded protein, which is still folding to a more advantageous low-energy shape. When the protein is fully folded, it separates from DnaK, which then binds to ATP once again, ending the cycle and setting the stage for the following round of protein folding. Any protein that may not have been folded completely is then taken over by the GroEL chaperonin system to complete the folding.

TABLE 1.2

Diseases Caused Due to Misfolding of Proteins

Name of Disease	Protein Affected
Alzheimer's disease	Misfolded β-amyloid proteins
Cystic fibrosis	Misfolded CFTR protein
Gaucher's disease	Misfolded β-glucocerebrosidase
Fabry disease	Misfolded α-galactosidase

Few of the examples of diseases caused due to protein misfolding are listed as follows (Figure 1.2):

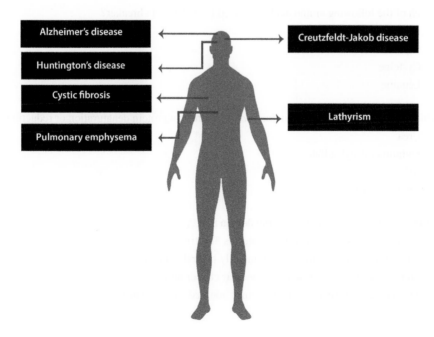

FIGURE 1.2 Diseases arise due to protein misfolding.

1.5 Conclusions

Amino acids play a significant role as building blocks of polypeptides in organisms. The amino acid composition of a given protein is influential in determining the characteristic features of the corresponding protein in question. The amino acid sequence or the primary structure of proteins, in other words, is responsible for determining the 3D structure of proteins. Proteins do not attain a functionally active state until they have undergone folding to reach a peculiar 3D structure. The different levels of structural organization of proteins, when understood vividly, can be helpful to extrapolate their functional significance to the cell or organism. Protein folding is a thermodynamically favorable process and occurs until a native and stable conformation is reached. Protein misfolding can be a consequence of mutations or wrong covalent interactions leading to serious downstream effects on the cell. Therefore, the role of individual amino acids, their special features, their peculiar arrangement to form a polypeptide, the peptide bond geometry, hierarchy of protein structural arrangement and protein folding and misfolding are all pivotal facets in proteomic studies.

REFERENCES

Baker, D. (2000). A surprising simplicity to protein folding. *Nature*, *405*(6782), 39–42. https://doi.org/10.1038/35011000

Shea, J. E., & Brooks, C. L. (2001). From folding theories to folding proteins: A review and assessment of simulation studies of protein folding and unfolding. *Annual Review of Physical Chemistry*, *52*, 499–535. https://doi.org/10.1146/annurev.physchem.52.1.499

Vendruscolo, M., Zurdo, J., MacPhee, C. E., & Dobson, C. M. (2003). Protein folding and misfolding: A paradigm of self-assembly and regulation in complex biological systems. *Philosophical Transactions. Series A, Mathematical, Physical, and Engineering Sciences*, *361*(1807), 1205–1222. https://doi.org/10.1098/rsta.2003.1194

Exercises 1.1

1. Which of the following amino acid is referred to as the helix breaker?
   a. Proline
   b. Glycine
   c. Cysteine
   d. Leucine

2. Which of the following chemical does *not* play a role in protein denaturation or reduction?
   a. Urea
   b. Sodium dodecyl sulfate
   c. APS
   d. Guanidine HCl

3. Anfinsen's experiment established the dogma that…?
   a. Protein's secondary structure dictates its 3D structure
   b. Protein refolding is a thermodynamically favorable process
   c. Protein folding requires aid of molecular chaperones
   d. The native conformation of a protein is adopted spontaneously

4. Ramachandran plot can be useful to study which level of protein structure?
   a. Primary
   b. Secondary
   c. Super secondary
   d. Tertiary

5. The 2 different forms that a given amino acid can adopt around the central chiral carbon atom are known as?
   a. Tautomers
   b. Enantiomers
   c. Both a and b
   d. None of the above

6. Turns and Loops are examples of which level of protein structure?
   a. Primary
   b. Secondary
   c. Super secondary
   d. Tertiary

7. Which of the following amino acid is most likely to be present on the outer surface of proteins?
   a. Lysine
   b. Valine
   c. Methionine
   d. Phenylalanine

8. The titration curve of glycine showed two pK_a values, i.e., 2.34 and 9.6. At which of the following pH value(s) will glycine show maximum buffering capacity?
   a. 2.34
   b. 9.6
   c. At both 2.34 and 9.6
   d. 5.97

9. Histones are highly alkaline proteins with very high pI (~10.8). Which amino acid residues would you expect in relatively large number in histones?
   a. Arginine
   b. Isoleucine
   c. Tyrosine
   d. Proline

10. At a pH of 12, what charged group(s) are present in isoleucine?
    a. $-NH_3^+$
    b. $-COO^-$
    c. $-NH_2^+$
    d. a and b

11. Following plot is derived from a method, which is used for the detection of secondary structures of proteins. Which method is it? Briefly describe its working principle.

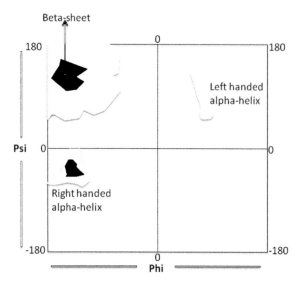

Answers

1. a
2. c
3. d
4. b
5. b
6. c
7. a
8. c
9. a
10. b
11. The plot illustrated above is a Ramachandran plot. This plot helps in understanding the relative distribution of helices and β-sheets of a protein based on psi and phi angles of the amino acids present in the proteins. By mapping the range of psi and phi values to different regions of the plot, one can distinguish different secondary structures. Ramachandran plot has "allowable" regions and "disallowable" regions. The polypeptides whose side chain and main chain atoms cause steric hindrance are represented in the disallowed region.

2

Protein Chemistry to Proteomics

Preamble

Proteins are the most dynamic and versatile macromolecules in a living cell, which regulate essential activities of the cell. The classical protein chemistry studies aimed at isolation, identification and functional elucidation of proteins. Protein chemistry deals mainly with sequence retrieval of the amino acids, followed by generating an X-ray crystallographic image of the same and finally elucidating the function of the protein. However, as the technology advanced, a new discipline "proteomics" came into existence, which aimed to look into the protein properties from a global perspective i.e., not undertaking one protein at a time but an entire set of proteins in the milieu. Proteomics mainly deals with the interaction of proteins amongst each other, thereby attempting to create a modeling strategy to answer the systems biology of an organism as a whole. The advancement in technologies like mass spectrometry and electrophoresis, the availability of genome sequences and bioinformatics tools are major contributors for the transition from protein chemistry to proteomics.

Terminology

- **Difference gel electrophoresis (DIGE):** An electrophoretic technique that allows more than one protein sample to be run simultaneously on a single 2-DE gel by carrying out differential labeling of each sample. This helps in eliminating any gel-to-gel variations and simplifies the process of analyzing proteomic alterations in two different conditions. An internal standard consisting of equal amount of all samples (in the experiment) is also run, further reducing any variation.
- **IPG strip:** Commercially available immobilized pH gradient (IPG) gel strips have replaced tube gels. They have considerably facilitated the process of isoelectric focusing by eliminating the tedious steps of gel preparation and pH gradient establishment using ampholyte solutions. These strips, available across the pH range, contain a preformed pH gradient immobilized on a precast polyacrylamide gel placed on a plastic support.
- **Protein microarrays:** These are miniaturized arrays normally made of glass, polyacrylamide gel pads or microwells, onto which small quantities of several proteins are simultaneously immobilized and analyzed. Protein microarrays can be generated by either traditional cell-based methods or more recently developed cell-free methods.
- **Ionization source:** This is responsible for converting analyte molecules into gas-phase ions in a vacuum.
- **Mass analyzer:** The mass analyzer resolves the ions produced by the ionization source on the basis of their mass-to-charge ratios.

2.1 Introduction

The central dogma of life places the translation process and protein products at the final stage (Figure 2.1). Proteins play a crucial part in cell growth and function. The DNA sequence determines the sequence of amino acids in the protein, but it is the modifications in the proteins resulting from alternative splicing and post-translational modifications (Figure 2.2), which finally dictates the physiological function of the

DOI: 10.1201/9781003098645-3

protein. Additionally, the structure and function of proteins are closely associated and their 3D structure governs their function. Hence, an approach should be available so that all information regarding proteins can be obtained in a high-throughput manner.

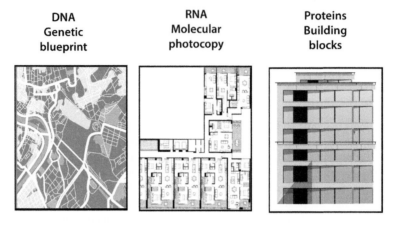

FIGURE 2.1   The central dogma; understanding protein function is key to biology.

Proteomics differs from the conventional protein chemistry approach of identification, structural and functional elucidation of proteins in many aspects. Proteomics aims to decipher the protein properties from a global perspective. This approach becomes a holistic model while studying the system because in a system, the protein in question is not alone. It is associated and interacting with several other proteins and biomolecules that together determine its role in the system. Protein chemistry and proteomics also differ in the techniques they employ as well as the scale of usability. While traditional protein chemistry also employs techniques like electrophoresis and mass spectrometry (MS), it is the advancement of these techniques along with genome sequence which led to the divergence from protein chemistry to proteomics. The sensitivity, resolvability, robustness and high throughput approaches made the transition possible. Expansion in the field of genome sequencing, bioinformatics and microarrays is also a major catalyst in the development of proteomics. This field is continuously progressing with the continuous evolution of several advanced technologies.

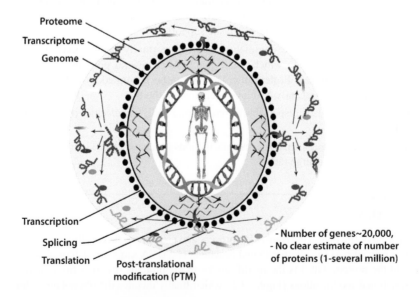

FIGURE 2.2   The complexity of the human proteome.

2.2 Evolution of Proteomics

The number of protein-coding genes in the human genome is approximately 5% of the entire genome. It is an astonishing fact that how this 5% of the genes account for the entire diversity of proteins in the cell. It is the dynamic properties of proteins which are responsible for all the functions in the human body, and any perturbation leads to several diseases. Proteomics emerged from protein chemistry solely because of the advancements in the existing technologies (Figure 2.3). In the following sections, a more detailed description of the advancements is explained.

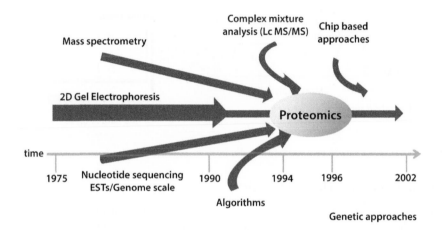

FIGURE 2.3 The emergence of proteomics.

2.2.1 Advancements in Mass Spectrometry

MS has been in existence for the last several decades, since the days of protein chemistry. MS ideally measures the mass of an analyte by producing charged molecular species in vacuum and their separation by magnetic and electric fields based on mass to charge (m/z) ratio. The prerequisite for a MS analysis is the generation of ions. However, proteins being large soluble polymers of amino acids could not be ionized by the conventional gas chromatography techniques without fragmenting them into constituent amino acids. This, for the time being, limited the usage of MS in protein chemistry. However, with major discoveries of soft ionization techniques like matrix-assisted laser desorption/ionization (MALDI) and electrospray ionization (ESI), MS became a robust analytical technique for the protein study.

With the advancement in ionization techniques, sophisticated mass analyzers and detectors, the mass analysis of as low as 1 ppm with excellent resolving power is possible. The sensitivity of the MS has also increased, making it possible to detect the 10^{-18} (attomole) mole of a sample. The tandem MS involving two mass analyzers like time of flight (TOF) and TOF/Q-TOF helps in protein sequencing. The Edman degradation process of amino acid sequencing though was extremely useful but was not a high-throughput approach as it could sequence a stretch of maximum of 40 amino acids, whereas MS could do it for large proteins. The ability of MS for relative and absolute quantitation of proteins by employing iTRAQ, ICAT, and SILAC labels has further advanced the field of quantitative proteomics.

Protein analysis by MS was challenging due to the complete degradation of samples owing to hard ionization techniques. This limitation was overcome by the development of soft ionization techniques such as MALDI and ESI. The vaporized sample is ionized by means of an electron beam in ESI or by a laser beam in MALDI. This results in charged peptide fragments, which are accelerated toward the mass analyzer. These two techniques greatly impacted proteomic studies as they facilitated MS analysis of protein samples. Protein sequencing by Edman degradation is time-consuming and cumbersome. Several rounds of sequencing are required for analysis of long polypeptide chains. Protein sequencing by MS, however, is much faster, which allows a large number of samples to be analyzed in the same amount of time (Figure 2.4).

MASS SPECTROMETRY

Multiple rounds of sequencing required to analyze long polypeptide
chains by Edman degradation. Sequencing by mass spectrometry is rapid
there by facilitating analysis of large number of samples.

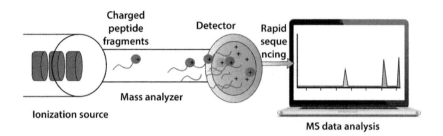

FIGURE 2.4   Mass spectrometric analysis vs. Edman degradation.

2.2.2 Advancements in Electrophoresis

Electrophoresis refers to the process of separation of charged particles under the influence of an electric field. Electrophoretic separation of proteins has been used widely for quite a long time. However, recent advances in electrophoretic techniques in the form of protein separation, staining, and detection have advanced this technique for further usage in proteomics.

The first advancement came in the transition from tube gels to immobilized pH gradient strips (Figure 2.5). Earlier, isoelectric focusing was performed using tube gels which offered poor stability as the pH would often change resulting in erroneous results. The tube gels showed a lot of variations and would often break after the addition of concentrated samples. Later, Prof. Angelika Gorg made a remarkable contribution to the field of electrophoresis by substituting the tube gels with immobilized pH gradient strips (Görg et al., 1995, 2000, 2009). These were acrylamide-coated plastic strips containing ampholytes of various pH spread across them. The biggest advantage of these strips was the stability of the pH ampholytes inside the gel. As a result, gel-to-gel variations got reduced and the physical stability of the strips was enhanced, as much as, they can be stored after the first dimension for a week at −20°C before the second dimension can be done.

Staining techniques also increased the usability of gel electrophoresis in proteomic studies. The conventional Coomassie brilliant blue dye had the limitation that it required 40 µg protein. Although silver staining increased the sensitivity, it had imitations in terms of background staining and incompatibility with MS. To overcome these problems, cyanine dyes were introduced for staining which has extreme sensitivity and specificity. The cyanine dyes interact with the protein through their hydroxysuccinamide residues and fluoresce when subjected to the light of appropriate wavelength. This property of cyanine dyes is utilized in developing a technique known as difference gel electrophoresis (DIGE). The biggest advantage of DIGE was its ability to resolve gel-to-gel variations. The sensitivity of the dyes coupled with the latest software could detect a small change in the expression level of a protein with reliability. Although electrophoretic techniques such as two-dimensional gel electrophoresis (2-DE) are now considered to be primitive in proteomics; however, quantitative approaches such as 2D-DIGE still hold a good place in proteomics.

2.2.3 Completion of Genome Sequence Projects and Introduction to Bioinformatics

The Human Genome Project, headed by six countries was completed in 2003 by 20 laboratories around the globe. The Human Genome Project estimated the number of coding genes in our genome to be roughly 25,000. If each gene has at least two splice variants and each protein has at least two post-translational modifications, the protein pool will have 1,00,000 proteins. One can imagine the huge data

A)

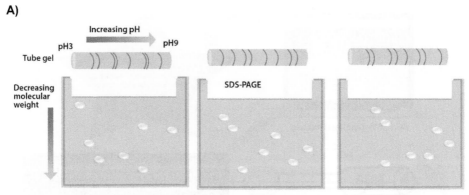

2-DE of same protein sample carried out using tube gels - variations in gel pattern

B)

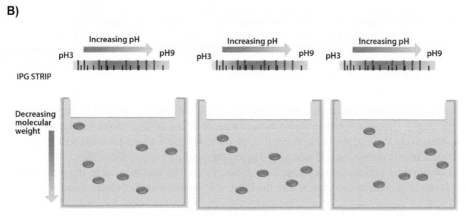

2-DE of same protein sample carried out using IPG strips - reproducibility in gel pattern

FIGURE 2.5 Two-dimensional gel electrophoresis using (A) tube gels and (B) IPG strips. The pH gradient in tube gels was established using ampholyte solutions, consisting of low molecular weight organic acids and bases that are subjected to an electric field. These gradients are not always very stable and tend to break down upon addition of concentrated samples. Analysis of the same protein mixture by 2-DE using tube gels resulted in a lot of variations across gels. The problem of reproducibility has been overcome to a large extent by the development of IPG strips, which are commercially manufactured gel strips having a preformed pH gradient. These strips only need to be rehydrated before use for 2-DE. Minimal gel-to-gel variations are observed when the same sample is run by 2-DE using IPG strips, thereby making them extremely suitable for large-scale proteomic applications.

that was generated from the Human Genome Project and many bioinformatics approaches were used to handle that. Bioinformatics led to the development of huge databases, which were designed to store information that could be accessible to any laboratory. Along with the Human Genome Project, many organisms' genomes also got sequenced under other genome projects. A combination of informaticians, biologists and chemists together led to the development of this entire branch of bioinformatics. The need for the development of extensive databases, software, and tools arose from the completion of the sequence projects of many organisms. The concept of reverse genetics and engineering also popularized with the availability of this genome sequence data. *De novo* sequencing of proteins using a mass-spectrometry-based approach led to the field of reverse genetics, where by knowing the sequence of the protein, one can elucidate the sequence of the genome (Figure 2.6).

2.2.4 Introduction to Protein Microarray

The success of DNA microarray also led to the development of a similar approach for protein microarrays. Like DNA microarray, the idea was to immobilize proteins in the chips or antibodies against

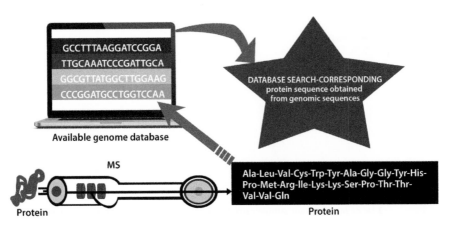

FIGURE 2.6 Completion of several genome sequence projects. The genomic DNA is cleaved using a suitable restriction endonuclease and inserted into the bacterial artificial chromosome. The amplified sequences are sequenced using an automated sequencer and then mapped by aligning the overlapping fragments to obtain the original DNA sequence. Genome sequences of several organisms, including humans, have been completed and these genome databases are extremely useful in correlating gene and protein sequences. Several databases are now readily available which can easily help in identifying the gene sequence of a protein that has been sequenced by MS.

the proteins on the chips and then detect them. However, this success proved to be short-lived and extremely costly. Unlike DNA microarray, where a piece of the DNA is immobilized on the chip, in protein microarray the entire protein needed to be immobilized with its function still intact, or else, if its 3D structure is destroyed it would also destroy the binding site for antibodies for detection. The printing of so many proteins is proved to be extremely cumbersome that led to the establishment of various forms of protein microarray, where the DNA was printed and the protein was translated in situ for detection.

2.3 Proteomics vs. Protein Chemistry

The debate persists whether proteomics emerged from protein chemistry or is it a new field altogether. Although proteomics employs the basic techniques and principles of protein chemistry, its approach and ultimate goals are different. While protein chemistry has always dealt with single protein, proteomics is more interested in studying the global proteome as a whole, making its usability for high-throughput studies. The approach of protein chemists has always been targeted; isolation of proteins followed by structural elucidation and functional analysis. On the other hand, proteomics researchers have always taken up the challenge of studying a large number of proteins together and are more interested in the interactions between the proteins, which ultimately helps to understand the function of the protein. In a way, this approach leads to long-term goal of understanding the role of the protein in the cell or the organism. The protein chemists have emphasized the structural aspects of the protein, while the proteomics researchers have aimed to look at the proteins in terms of developing a mathematical model for the system as a whole, using a systems biology approach.

Protein chemistry began its sequence retrieval using the Edman degradation technique, whereas proteomics does the same using MS. Edman degradation method still holds good but only when there is a limited number of proteins; however, for high throughput studies, the mass spectrometry-based sequencing has become the method of choice. The additional benefit of mass spectrometry-based sequencing is that there is no need for sequencing the entire protein. Few peptide sequences and bioinformatics tools are sufficient to identify the protein. In summary, protein chemistry and proteomics have a unique outlook of the problem and methodology employed to address the problems; however, both are complementary to each other. An overview of various aspects of the advancements of proteomics is given in Figure 2.7.

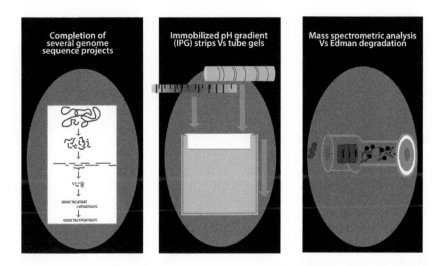

FIGURE 2.7   The advancements of proteomics with respect to protein chemistry.

REFERENCES

Görg, A., Boguth, G., Obermaier, C., Posch, A., & Weiss, W. (1995). Two-dimensional polyacrylamide gel electrophoresis with immobilized pH gradients in the first dimension (IPG-Dalt): The state of the art and the controversy of vertical versus horizontal systems. *Electrophoresis, 16*(7), 1079–1086. https://doi.org/10.1002/elps.11501601183

Görg, A., Drews, O., Lück, C., Weiland, F., & Weiss, W. (2009). 2-DE with IPGs. *Electrophoresis, 30*(Suppl 1), S122–S132. https://doi.org/10.1002/elps.200900051

Görg, A., Obermaier, C., Boguth, G., Harder, A., Scheibe, B., Wildgruber, R., & Weiss, W. (2000). The current state of two-dimensional electrophoresis with immobilized pH gradients. *Electrophoresis, 21*(6), 1037–1053. https://doi.org/10.1002/(SICI)1522-2683(20000401)21:6<1037::AID-ELPS1037>3.0.CO;2-V

Exercises 2.1

1. The net charge of a protein when it comes to rest on an IEF gel is?
   a. Positive
   b. Neutral
   c. Negative
   d. Doubly positive

2. Which of the following is *not* a mass analyzer?
   a. TOF
   b. ESI
   c. Quadrupole
   d. Ion trap

3. What pH range can be used to separate a crude protein extract containing 100 proteins of varying isoelectric points?
   a. pH 2–3
   b. pH 4–5
   c. pH 8–9
   d. pH 3–11

4. In DIGE _____ dyes are used?
   a. Silver
   b. Coomassie
   c. Cyanine
   d. Pro-Q Diamond

5. *De novo* protein sequencing is done when?
   a. Protein structure is not known
   b. DNA sequence is not known
   c. Both a and b
   d. None of the above

6. What is the advantage(s) of 2D-DIGE over conventional 2-DE?
   a. Reduced gel-to-gel variation
   b. High reproducibility
   c. High sensitivity
   d. All of the above

7. The completion of Human Genome Project revealed the total number of protein coding genes
to be around 25,000. However, the number of human proteins is estimated to be in millions.
What could be the most probable reason for this discrepancy?
   a. Alternative splicing and post-translational modifications
   b. Presence of extremely large number of introns in the human DNA
   c. Transcription is an extremely slow process
   d. None of the above

8. Which of the following statement(s) is *not* true about 2-DE using tube gels?
   a. pH gradients are established by ampholyte solutions
   b. Gel-to-gel reproducibility is extremely good
   c. They often break down upon the addition of concentrated samples
   d. All of the above

9. N-terminal sequence of proteins can be determined using which of the following method?
   a. Sanger method
   b. Edman reaction
   c. Mass spectrometry
   d. All of the above

10. Which of the following statements is true for proteomics?
    a. It is a branch of plant genomics
    b. It is the study of the entire set of proteins expressed by any organism
    c. It is the study of the entire set of proteins expressed by humans
    d. It is the study of molecular components, including ribosomes, needed to synthesize proteins

11. You want to sequence a polypeptide, which has approximately 100 amino acids. You don't have the information about the free N-terminal. Out of the two methods, Edman degradation technique and mass spectrometry-based proteomics, which method would be more suitable in this scenario?

Answers

1. b
2. b
3. d
4. c
5. b
6. d
7. a
8. b
9. d
10. b
11. The Edman degradation technique and mass spectrometry-based proteomics both can be used for sequencing polypeptides. However, the polypeptide of interest, in this case, is of approximately 100 amino acids long. The Edman degradation technique is not ideal for sequencing the stretches longer than 40 amino acids. Whereas, polypeptides of up to 100 amino acids length can be analyzed using mass spectrometry. Besides, Edman degradation technique can only be used for free N-terminal proteins and in this case, we are not sure about the free N-terminal status of the protein. Based on the conditions mentioned above, the mass spectrometry-based proteomics would be preferred in this case.

Module II

Gel-Based Proteomics Techniques

3

Gel-Based Proteomics

Preamble

Gel-based proteomics is one of the versatile fields of proteomics, which has provided us with tools that can be used for protein separation, characterization as well as quantification. The goal of proteomic studies is to detect altered protein expression and modifications associated with a disease or to find molecular targets for biomarkers and therapy (Lescuyer et al., 2007). Gel-based proteomics includes techniques like one-dimensional sodium dodecyl sulfate-polyacrylamide gel electrophoresis (SDS-PAGE) or native-PAGE, two-dimensional gel electrophoresis (2-DE), and difference gel electrophoresis (DIGE). The proteins that are differentially expressed can be further identified using techniques such as mass spectrometry. Gel-based proteomics mainly exploits the principle of electrophoresis and provides information about protein properties such as molecular weight and isoelectric point. In this chapter, we will provide an overview of the different gel-based techniques and their working principles.

Terminology

- **Gel-based proteomics:** High-throughput techniques for separation of proteins from complex mixtures using polyacrylamide gels.
- **Electrophoresis:** The separation of charged molecules under the influence of an applied electric field.
- **IEF:** Isoelectric focusing (IEF) is the separation of proteins based on their isoelectric points (pI).
- **IPG:** Immobilized pH gradient
- **SDS-PAGE:** Sodium dodecyl sulfate polyacrylamide gel electrophoresis, which brings about separation based on their relative molecular mass.
- **2-DE:** Two-dimensional gel electrophoresis that carries out separation using isoelectric focusing in the first dimension, followed by SDS-PAGE in the second dimension.
- **2D-DIGE:** 2D difference gel electrophoresis is an advanced form of 2-DE that allows simultaneous analysis of test and control samples on a single gel by carrying out differential labeling of each sample. This minimizes gel-to-gel variations and enables easy processing of a large number of samples
- **ANOVA:** Analysis of variance is a collection of statistical models and their associated estimation procedures (such as the "variation" among and between groups) used to analyze the differences among group means in a sample.

TABLE FOR HISTORY AND EVOLUTION

- **1930:** Arne Wilhelm and Kaurin Tiselius invented Electrophoresis.
- **1975:** 2-DE was reported independently by Klose and O'Farrell.
- **1990s:** Angelika Gorg contributed for development of IPG strips.

3.1 Electrophoresis

Electrophoresis is the widely used technique for protein separation and works on the principle of migration of charged molecules in a gel matrix towards the oppositely charged electrode, under the influence of an electric field. It is a powerful technique for finer protein separation and visualization of these separated proteins. Electrophoresis was invented by Prof. Tiselius in 1930 as the moving boundary method to study the electrophoresis of proteins (Longsworth & MacInnes, 1939). It has been extensively used since then and numerous advancements have been brought in this technique.

3.2 One-Dimensional Gel Electrophoresis

One-dimensional gel electrophoresis (1-DE) relies on the principle of separation of protein molecules based on their charge to mass ratio or molecular weight. The low molecular weight proteins are able to migrate to a larger distances on the gel as compared to the higher molecular weight proteins (Gallagher, 2012). Polyacrylamide gels are formed from the polymerization of two compounds; acrylamide and N,N′-methylenebisacrylamide (the crosslinking agent). The polymerization is initiated by the addition of ammonium persulfate (APS) along with either 3-(dimethylamino)propionitrile (DMAP) or N,N,N′,N′-tetramethylethylenediamine (TEMED). The commonly employed 1-DE techniques include sodium dodecyl sulfate-poly acrylamide gel electrophoresis (SDS-PAGE) and Blue Native (BN) PAGE.

3.2.1 Sodium Dodecyl Sulfate-Poly Acrylamide Gel Electrophoresis

It is a technique that uses an anionic detergent namely sodium dodecyl sulfate, which provides a uniform negative charge on protein molecules (Figure 3.1). This entails the electrophoretic separation to be brought about only on the basis of the molecular weight of the proteins (Brunelle & Green, 2014).

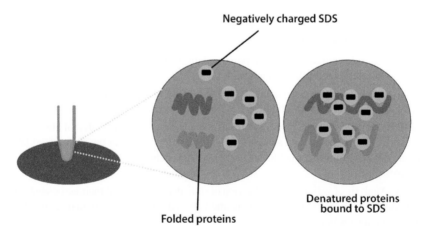

FIGURE 3.1 The uniform charge to mass ratio of proteins upon binding to SDS. After addition of β-mercaptoethanol to the sample, it acts as a reducing agent and breaks the disulfide bonds and unfolds the protein, SDS provides a uniform negative charge to denatured protein.

Electrophoresis is a powerful technique for protein separation and separated proteins can be visualized after subsequent staining steps. It is based on the principle of migration of charged proteins in an electric field. The polyacrylamide gel containing SDS is cast between glass plates arranged as a vertical slab in the same buffer that is used for electrophoresis. The molecular dimensions of the pores can be controlled by varying the amount of N, N′-methylenebisacrylamide with free-radical crosslinking being facilitated by Ammonium persulfate (APS) and Tetramethylethylenediamine (TEMED).

For a discontinuous gel, two gel solutions are prepared; resolving and stacking gel solutions. Firstly resolving gel solution is poured to a particular level. It is followed by pouring of stacking gel solution after the resolving gel is solidified. Sample wells of uniform size, shape and separation are made using a comb, which is placed in the gel as soon as stacking gel solution has been poured. After the gel has polymerized, the comb is removed, providing a gel ready for the process. SDS is a negatively charged anionic detergent that binds to protein molecules and causes them to denature. The dithiothreitol (DTT) breaks any disulfide linkages that may be present. The binding of SDS causes the proteins to have a uniform charge-to-mass ratio, thereby allowing separation purely on the basis of molecular weight.

The protein samples are loaded into the wells with the help of a micropipette. Once the samples have been loaded, a direct current supply between 90 to 250 V is passed depending upon the size of the gel for a time sufficient to separate the protein mixture into discrete bands based on their molecular weight (Figure 3.2). The progress of electrophoresis can be observed with the help of tracking dye. The larger proteins are retarded in the gel and remain close to the point of application while the smaller proteins migrate further along the gel. The gel is then stained with either coomassie or silver staining and viewed to observe the various discrete protein bands.

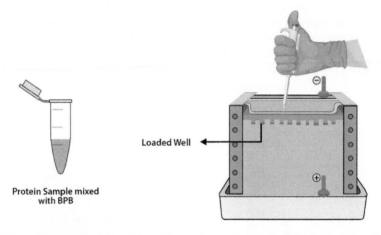

Loaded Well ◄────

**Protein Sample mixed
with BPB**

FIGURE 3.2 SDS-PAGE electrophoresis technique. The protein samples are mixed with the dye bromophenol blue (BPB) and loaded into the electrophoresis unit.

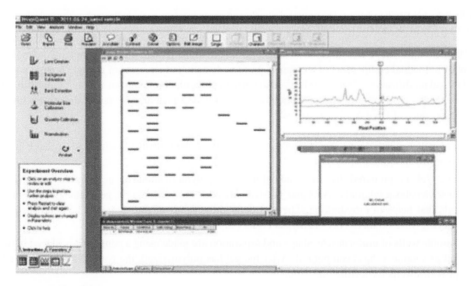

FIGURE 3.3 SDS-PAGE data analysis using ImageQuant TL software. SDS-PAGE analysis is done to study the expression of a protein from the control and disease sample, to detect the molecular weight of the protein using the molecular weight marker, and to detect the quantity by the intensity of the protein. For this purpose, specialized software can be used.

The steps for SDS-PAGE data analysis using ImageQuant TL software (Figure 3.3) are as follows:

- Load the gel for analysis into the software. Specify the number of lanes and lane width. Drag the image, the lanes will be created automatically.
- Once the parameters are set, click "next".
- Specify the molecular weight of the standard. Now click on the marker to know the molecular weight of the sample in the gel.
- Specify the unit in which the quantity of the protein has to be displayed. Click on the band to know the quantification value of the protein band.

3.2.2 Blue Native PAGE

BN-PAGE allows proteins to be studied in their native state and does not involve any denaturation step. The Coomassie dye used herein provides the necessary charge to the protein complexes and further helps in their separation during migration on the gel. Thus, the electrophoretic mobility mainly relies on the negative charge that is imparted by the dye, and the size and shape of the protein complexes that are being studied. It can be used for the identification of multi-protein complexes and hence provides an integrative view of the protein, as they need to be separated in the native conditions (Figure 3.4) (Nijtmans et al., 2002).

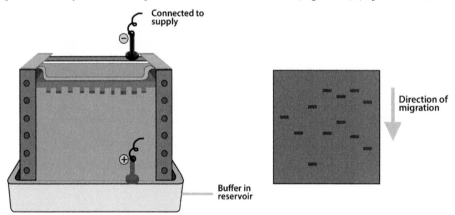

FIGURE 3.4   BN-PAGE technique. The vertical electrophoresis apparatus showing the buffer tank, with a cathode (negative terminal) and an anode (positive terminal). The migration of proteins occurs from a negative to a positive direction.

1-DE technique can be effectively used to determine the whole protein molecular weight, the molecular weight of individual components of a complex protein, detection of isoforms of a single protein and post-translational modifications. It can also be used to purify a particular protein from a complex mixture, which has to be further used for different applications and can provide an integrative view of the protein function.

The procedure of BN-PAGE is as follows:

- The polyacrylamide gel is cast between vertical glass plates using the same buffer that is used for electrophoresis.
- The gel is prepared by free radical-induced polymerization of acrylamide and N,N′-methylenebisacrylamide. APS and TEMED are added to facilitate the generation of free radicals that helps in crosslinking. The molecular dimensions of the pores can be controlled by varying the amount of N,N′-methylenebisacrylamide.
- Sample wells of uniform size, shape and separation are made using a comb that is placed in the gel as soon as it has been poured. After the gel has polymerized, the comb is removed which provides wells for loading the samples.
- The protein sample present in a suitable buffer system is mixed with the coomassie blue dye, which provides the necessary charge to the protein complexes thereby facilitating their separation in the gel. Unlike SDS, the dye does not denature the proteins but binds to them in their native state.

- The protein samples are then loaded into the wells with the help of a micropipette. Once the samples have been loaded, a direct current supply of around 100–350 V is passed depending upon the size of the gel for a time sufficient to separate the protein mixture into discrete bands based on their charge-to-mass ratio.

- Progress of electrophoresis can be observed with the help of tracking dye. The larger proteins are retarded in the gel and remain close to the point of application, while the smaller proteins migrate further along the gel. The gel is then stained with Coomassie Brilliant Blue and viewed to observe the various discrete protein bands.

The answers to the below questions will make the concept of the electrophoretic technique clear. Figures 3.5 and 3.6 show an activity to comparatively analyze SDS-PAGE and BN-PAGE.

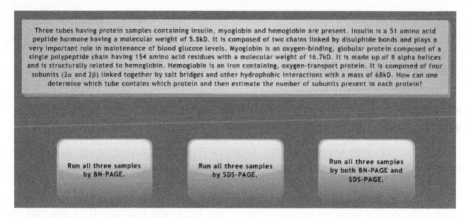

FIGURE 3.5 SDS-PAGE vs. BN-PAGE (interactivity 1).

1. Three tubes having protein samples containing insulin, myoglobin, and hemoglobin are to be tested. Insulin is a 51 amino acid peptide hormone having a molecular weight of 5.8 kDa. It is composed of two chains linked by disulfide bonds and plays a very important role in the maintenance of blood glucose levels. Myoglobin is an oxygen-binding, globular protein composed of a single polypeptide chain having 154 amino acid residues with a molecular weight of 16.7 kDa. It is made up of eight alpha helices and is structurally related to hemoglobin. Hemoglobin is an iron-containing, oxygen-transport protein. It is composed of four subunits (2a and 2b) linked together by salt bridges and other hydrophobic interactions with a mass of 68 kDa. How can one determine which tube contains which protein and then estimate the number of subunits present in each protein?

   A. Run all three samples by BN-PAGE.
   B. Run all three samples by SDS-PAGE.
   C. Run all three samples by both BN-PAGE and SDS-PAGE.

   A comparison of the electrophoretic separation profiles of hemoglobin, insulin, and myoglobin using both BN-PAGE and SDS-PAGE will allow one to determine which tube contains which protein sample and how many subunits may be present in each protein. In the case of BN-PAGE, since the molecular weight of each protein is known, the migration distance will indicate which band corresponds to which protein. In SDS-PAGE, more than one band will appear for hemoglobin (Hb) and insulin indicating that they have multiple subunits. The thicker bands are indicative of more than one subunit of nearly identical molecular mass.

2. Which technique would be suitable for the determination of the molecular weight of the enzyme α-amylase? BN-PAGE or SDS-PAGE?

   SDS-PAGE is commonly used for the determination of the molecular weight of an unknown protein by running it along with other protein markers of known molecular weight. Mobility of the unknown protein can therefore be used to determine its molecular weight from a standard graph.

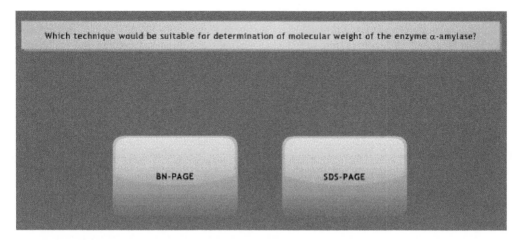

FIGURE 3.6   SDS-PAGE vs. BN-PAGE (interactivity 2).

3.3 Two-Dimensional Gel Electrophoresis

Two-dimensional gel electrophoresis (2-DE) separates proteins based on isoelectric point (pI) and molecular weight. It can be effectively used for applications such as differential proteomic analysis, isoforms separation or analysis of post-translational modifications (Gygi et al., 2000; Rabilloud, 2002). In 2-DE, the first separation occurs based on the pI of the proteins, and the second separation carried out orthogonally, occurs based on the molecular weight. Isoelectric focusing (IEF) separates proteins based on their pI and results in immobilization of the protein at its pI at which the overall charge on the protein is zero. The immobilized pH gradient (IPG) strips are equilibrated with DTT and iodoacetamide (IAA) for reduction and alkylation, respectively. This makes the strip ready for the second-dimensional run. In the second dimension, separation based on size occurs by SDS-PAGE. The gels are stained with coomassie blue dye to stain the protein spots. Destaining follows this step and then the gels are scanned. The gel images thus obtained are analyzed and significant spots are excised and digested for further analysis by mass spectrometry (Figure 3.7).

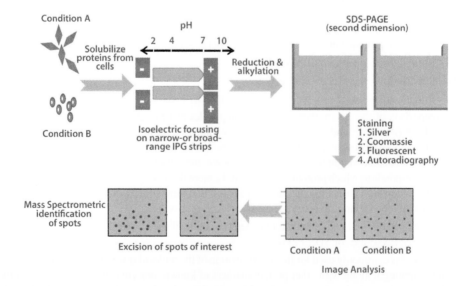

FIGURE 3.7   An overview of two-dimensional gel electrophoresis (2-DE).

The procedure of 2-DE is as follows:

- Prior to IEF in 2-DE, the commercially available IPG strips must be rehydrated. This is done by soaking them for 10–20 hrs in the protein sample, which has a suitable buffer solution. Once the strips are rehydrated, they are covered with mineral oil to prevent evaporation of solution and drying of the gel.
- Once the IPG strips have been rehydrated, the strip is placed in the tray and the sample is added to the sample cup followed by the cover fluid to prevent drying of gel. In passive loading, the gel strip is placed face down in the cover fluid containing the sample for 10–12 hrs.
- **IEF:** These loaded strips are then focused on an IEF unit by passing current. The various proteins of the sample mixture migrate in the electric field and come to rest when the pH is equal to their pI, i.e., they become neutral and are no longer affected by the electric field. Progress of electrophoresis is monitored by means of a tracking dye like bromophenol blue (BPB), which is a small molecule and therefore migrates ahead of all other proteins.
- **SDS-PAGE:** The IPG strip is equilibrated in a reducing agent like DTT followed by an alkylating agent, IAA, which prevents reformation of the reduced bonds. This strip containing the separated proteins is then placed on the SDS-polyacrylamide gel slab for further protein separation in the second dimension based on their molecular weight.
- The proteins on the IPG strip are then subjected to SDS-PAGE by applying a direct current between 100–350 V depending upon the size of the gel. Proteins that if present as a single band on the IPG strip due to similar pIs can now be separated based on their molecular weight with smaller proteins migrating farther.
- View of a sample gel, which has been run by 2-DE and stained with coomassie blue. Each spot provides information about the molecular weight (MW) and pI of the proteins.

Passive rehydration: It is the process of allowing the IPG strip to swell with protein sample without applying any current (Figure 3.8).

2-DE, Isoelectric Focusing
IPG Strip rehydration-Passive rehydration, 10-20h

Mineral Oil

FIGURE 3.8 Passive rehydration. The process of allowing the IPG strip to swell with protein sample without applying any current.

- Prepare the sample for loading on the strip equivalent to 600 µg protein for 18 cm IPG strip. In case the user is loading for 7 and 24 cm, the calculation can be done accordingly.
- Load the sample in the reswelling tray/manifold and avoid air bubbles.
- Remove the strip from −20°C and allow it to thaw.
- Place the strip on the reswelling tray with the gel side facing down and keep it for half an hour.

- Add cover fluid to the strip to prevent evaporation. While adding avoid air bubbles. The strip must be covered with oil.
- The overnight process helps rehydrate the IPG strip with a protein sample in the absence of an electric field. Once the passive rehydration is over, the strip is ready for IEF run.

Active rehydration: In active rehydration, the sample is loaded in a single electrode manifold unit in case of a single strip or strip loaded into manifold directly if many strips are there. Unlike the use of a rehydration tray in passive rehydration, for active rehydration manifold is used (Figure 3.9).

2-DE, Isoelectric Focusing
IPG Strip rehydration-Passive rehydration, 10-20h

Cover Fluid (to prevent drying of gel)

FIGURE 3.9   Active rehydration. The process of allowing the IPG strip to swell with protein sample with applying current.

- The electrode assembly is held together and placed at the correct position by the plastic support. The system goes in for automatic calibration for display, buzzer and light to check for working condition.
- Connect the instrument properly. Switch ON the system and select active rehydration protocol for 5 hrs at 20 V/50 mA for each strip. Subsequently, user proceeds for the IEF step.

Isoelectric focusing: Proteins exhibit unique isoelectric property and these properties are exploited for the separation of individual proteins from a pool of proteomes (Figure 3.10).

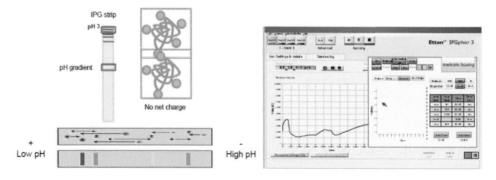

FIGURE 3.10   Isoelectric focusing. The separation of proteins on the basis of their net charge.

A representative protocol is demonstrated as following:

- Place the manifold on the table. Take a clean tissue paper and clean all the lanes of the manifold to make it free from dust and dry it completely.

- Place the instrument on a leveled surface. Take the manifold and position it inside the groove of the instrument. Ensure that the surface is leveled by placing the round spirit level that can be adjusted if necessary.
- Pick the strip end with forceps from the reswelling tray (which is kept for passive rehydration). Drain out excess oil by tapping it on tissue paper without folding the strip and keep it straight. Place the positive side of the strip on the positive end of the instrument with the gel side up.
- Cut the paper wicks into two pieces. Add distilled water to the wicks to get wet. Place it on the strip end such that one end of the wick overlaps the end of the gel on the strip. The paper wicks absorb impurities/salts and aid in current conductance.
- Add cover fluid to the well containing the strips and to the other wells. Make sure that the level of cover fluid must not overflow.
- Place the electrodes on either side of the tray such that the tip of the electrode touches the paper wick and is immersed in the cover fluid, if not add some more oil and close the lid.
- Connect the instrument properly. Switch ON the system and start the software by clicking on the icon. Select the specific protocol for the sample to run the IEF. Users can change the protocol according to their requirements. User can change the select strip length and define the number of strips running in the experiment.
- The set protocol is reached when the sample preparation is perfect and the IEF can run without any interference. The set protocol will not reach if the sample preparation is imperfect having impurities and may interfere during the run.
- In case of an improper run, replace the wicks with the new ones and restart the run. The run will continue once the wicks are changed.
- Protein stops moving at a point when the net charge is zero which is known as the pI. Stop the run once the set voltage is reached and the run is completed.
- Remove the IPG strip carefully without bending. Store the strip at −80°C until further processing. Once the equilibration buffers are prepared, the strip can be taken out to carry out the SDS-PAGE.

Equilibration: Separated proteins in the strip undergo an equilibration step so that the multi-subunit proteins can be separated (Figure 3.11). This step is helpful for the prevention of reunion of the separated proteins and stabilization of the separated protein in the gel.

FIGURE 3.11 Equilibration of IPG strips.

Equilibration buffer-I and buffer-II are required for this step. **Equilibration buffer-I** consists of urea as denaturing agent, SDS to provide a uniform negative charge to the protein, DTT as the reducing agent of disulfide bond, glycerol as the stabilizing agent of polyacrylamide gel. **Equilibration buffer-II** consists of urea as denaturing agent, SDS to provide a uniform negative charge to the protein, IAA as the acetylating agent of reduced disulfide bond, glycerol as the stabilizing agent of polyacrylamide gel. A representative protocol is as follows:

- Place the IPG strip in the equilibration tray with the gel side up. Add the equilibration buffer-I to IPG strip and keep it on the shaker for 15 min. The strip must be covered with the buffer. DTT in equilibration buffer reduces the disulfide bond and helps to maintain all proteins in their fully reduced state. Remove the strips from the well and place them in the new well.
- Add equilibration buffer-II to the strip and keep it on the shaker for 15 min. IAA acetylates the reduced disulfide bond and prevents its oxidation, back folding and aggregating of protein subunits. Remove the strips from the well and place them in the new well. Wash the strips with tank buffer after the second equilibration step. Now the strip is ready for the SDS-PAGE run.

Second-dimension electrophoresis: Proteins exhibit different molecular weights depending on the amino acid composition. This property is exploited to separate proteins in second dimensions on SDS-PAGE gels (Figure 3.12).

2-DE, SDS-PAGE

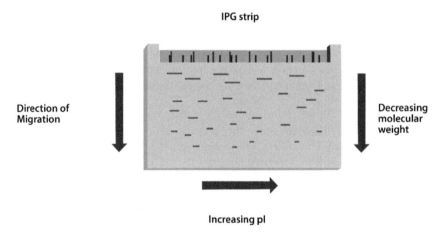

FIGURE 3.12   Second-dimension separation.

A representative protocol for second dimension electrophoresis is as follows.

- Prepare the gel casting template unit. Space within the glass plates helps to hold the gel or gel is formed within this space. For easy separation of glass plates, plastic sheets are used.
- Prepare 12.5% resolving gel cocktail consisting of 49 mL acrylamide and bisacrylamide, 30 mL 1.5 M Tris-HCl (pH 8.8), 1.2 mL 10% SDS, 1.2 mL 10% APS, 165 µL TEMED and 120 mL water. Add APS at the last minute. Pour the solution at once and avoid air bubbles.
- Spray 0.1% SDS to prevent drying off the gel, and help to produce an even surface on top of the gel. When the gel is becoming polymerized, start the process of equilibration for IPG strips.
- The tank buffer consists of a tris base, glycine provides proper conductance, and SDS adds a negative charge to protein for separation.
- Wash the strips with tank buffer after the second equilibration step. Now the strip is ready for the 2D run. Transfer the glass plate with gel slab into the glass plate holder to keep IPG strips.

- Place the strip in the gel and ensure that there shouldn't be any gap or air bubbles between the gel and the strip.
- Seal the strip and the gel using agarose sealing solution and wait until it solidifies. Agarose needs to be liquid hot before use.
- Place the glass plate containing strip and dummy plates in the gel holding unit. All the groves in the unit need to be filled. In case if the user is running only two gels, place each gel unit on either side of the unit. It helps for the even flow of voltage across the unit.
- Prepare 1× SDS buffer from 10×. Pour the buffer into the chamber till the maximum level (without the gel cassette) in the lower tank is achieved. The lid acts as an upper reservoir for the buffer chamber. Once the buffer is added to the maximum level, the unit is ready for 2DE run.
- Connect the setup, if not properly connected the power back will show an error. Set the voltage to 100 V for 1 hr and 300 V for 4 hrs. The water bath temperature to be set at 20°C as the 2DE separation is an exothermic reaction. To remove the excess heat, water is circulated inside the unit.
- SDS denatures the proteins and provides a negative charge to the proteins which allows the proteins to move from a negative to positive terminal based on the pore size of the gel and molecular weight of the protein.
- Place the gel in the distilled water to remove excess SDS that may interfere with staining. Now the gel is ready for the staining process.

3.3.1 Staining Methods

Different staining methods are used to stain the 1D or 2D gels. Some of the commonly used stains for 2D gels (Figure 3.13) are discussed below.

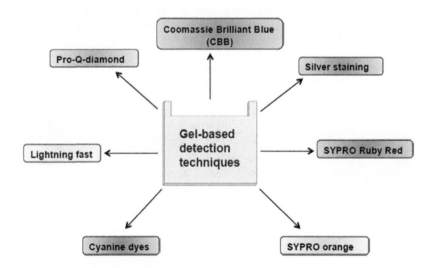

FIGURE 3.13 Commonly used stains to visualize 2D gels.

1. **Coomassie Brilliant Blue (CBB):** Coomassie blue dyes (R-250 and G-250) are low-cost, organic dyes that are easy to use for staining of proteins that have been separated by electrophoretic techniques (Figure 3.14). Gels are soaked in the dye solution dissolved in methanol and acetic acid. The excess stain is then washed off with a destaining solution. The higher affinity of proteins toward the dye molecules allows the protein bands to be selectively stained with the sensitivity of 8–100 ng without significant staining of the background. These dyes are also compatible with further MS-based applications.

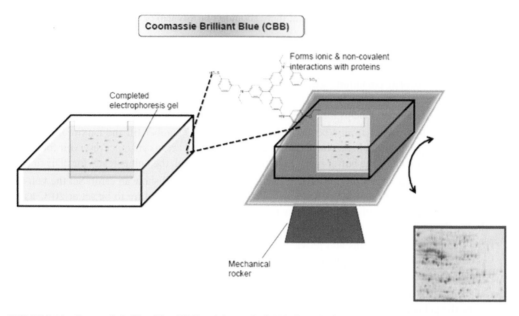

FIGURE 3.14   Coomassie Brilliant Blue (CBB) staining method. CBB forms ionic and non-covalent interactions with proteins.

2. **Silver staining:** Gels are saturated with silver ions in the form of either silver nitrate or ammonia-silver complex after fixing the proteins in the gel (Figure 3.15). The less tightly bound metal ions are subsequently washed off and the protein-bound silver ions are reduced to metallic silver using formaldehyde under alkaline conditions in the presence of sodium carbonate or citrate buffer solution. Although as little as 1 ng of protein can be detected by silver staining, the gel-to-gel reproducibility remains an issue. Compatibility of silver stains with MS is another issue, which has been overcome in recently introduced silver stains.

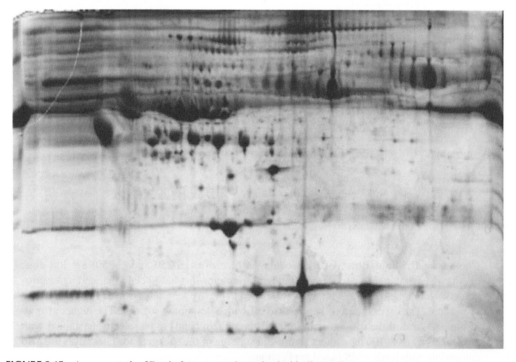

FIGURE 3.15   A representative 2D gel of serum proteins stained with silver stain.

3. **SYPRO Ruby Red:** This is a ruthenium-based metal chelate fluorescent stain that provides a single-step protein staining procedure with low background staining in polyacrylamide gels. They have been observed to be as sensitive as the silver stains (0.25–1 ng) with the linear dynamic range extending over three orders of magnitude, thereby showing better performance than CBB and silver stains. This stain can also be combined with other dyes thereby allowing multiple detections in a single gel.

4. **SYPRO orange:** This dye is less sensitive than SYPRO Ruby Red but is also capable of detecting proteins in SDS-PAGE gels in a rapid single-step process without the requirement for any destaining procedure. As little as 4–30 ng of protein can be detected by this fluorescent dye and it is compatible with MS-based applications. Two other similar dyes having comparable sensitivities and similar excitation and emission wavelengths are SYPRO Red and Tangerine.

5. **Cyanine dyes:** They are water-soluble derivatives of N-hydroxysuccinimide that covalently bind the ε-amino groups of a protein's lysine residues and are spectrally resolvable as they fluoresce at distinct wavelengths. The labeled protein samples can therefore be mixed and run on a single gel, thus eliminating the problem of gel-to-gel variations. Cy3, Cy5, and Cy2 having sensitivities of 0.1–2 ng are most commonly used for proteomic and MS-based applications.

6. **Lightning fast/deep purple:** A fluorescence-based stain obtained from the fungus *Epicoccum nigrum*, can be used for detecting proteins in 1-DE and 2-DE with sensitivity as low as 100 pg of protein. Stained proteins are excited by near-UV or visible light with maximum fluorescence emission occurring at around 610 nm. These dyes are suited for further use with Edman or MS applications.

7. **Pro-Q Diamond:** This fluorescent dye is capable of detecting modified proteins that are phosphorylated at serine, threonine, or tyrosine residues. They are suitable for use with electrophoretic techniques or with protein microarrays and offer sensitivity down to few ng levels, depending upon the format in which they are used. This dye can also be combined with other staining procedures thereby allowing more than one detection protocol on a single gel.

3.3.2 Gel Scanning and Analysis and Spot Picking

Gel scanning and Analysis: The proteome obtained by the 2-DE process is analyzed to identify the significant spots by comparing the control and the treated samples (Figure 3.16). It offers a flexible solution for the comprehensive visualization, exploration, and analysis of 2D gel data.

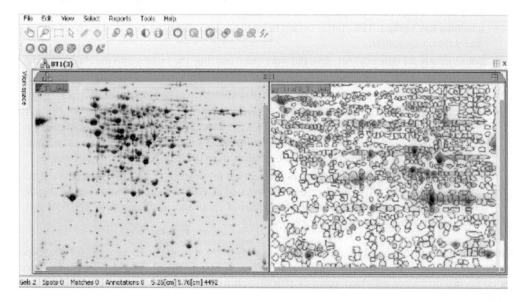

FIGURE 3.16 Methodology for gel scanning and 2D data analysis.

A representative protocol for gel scanning and analysis is as follows:

- **Gel scanning:** Place the gel in the middle of the scanner platform, and avoid air bubbles. Click on the shortcut icon or you can start the Scanner Control software using the Start menu. The Scanner Control window appears. The preview window appears. The user needs to set the parameters before going for scanning. During scanning, check the image in the Image Quant Preview window for saturation. Saturated pixels appear in red. Users can adjust the exposure time based on the image preview. After scanning with the optimum amount of exposure, label the gel with the proper annotations as project name, type of sample, etc.

- **Create the project:** Load the control and treated gel images in a folder for analysis. Once the gels are loaded, cropping needs to be done. Images rotating is done to align all the gels in the same orientation. Image processing does not affect original image files. The bold horizontal grid line helps to visualize the rotation. The flip icon is used when gel images are scanned in the wrong direction. The image can be flipped horizontally or vertically to produce their correct mirror image.

- Gel images can be cropped with the crop tool. This creates the gel images that only contain the selected area. Once the gels are cropped, the images are ready for analysis.

- Once gels have been added to a project, spot detection can be done. A spot delineates a small region in the gel where protein is present. Optimum spot detection parameters are defined and entered into the software. This shape is automatically differentiated by a spot detection algorithm and quantified in which its intensity, area and volume are computed. The parameter smooth helps to detect all real spots and split the overlapping ones. Subsequently, saliency and min area values helps to filter the noise.

- After spot detection, zoom each and every spot to check for a real spot. The software detects dusts and artifacts which should be removed from analysis.

- If the user finds the unwanted spot, he can delete it. If some spots are exactly at the same position across the gels, such spots can be landmarked that helps in matching. Once land marking is over, save the gels and import the gels into the matching folder.

- Matching of gels must done after land marking. Matching algorithm first matches the landmarked spots, then matches the nearby spots. In case if a smaller number of matches are produced, create few more landmarks and try again. Land marking must be performed in such a way that it should cover the entire gel area.

- Vector lines help to check the matching process i.e., how correctly the gels have matched with each other. If the vector lines are of the same length and in the same direction, we can say that matching has worked. The overlay option helps the user to check the profile of each gel and to detect matching pattern.

- Subsequently for data analysis, save the workspace and drag the images into the classes folder. Repeat the procedure until the matching is done and the user is satisfied with the end result.

- Spots across the gel can be selected and compared to know their pI, volume, intensity and MW etc.

- The protein spots can be represented in the 3D form with peak height denoting its intensity and can be rotated to view from different angles. This helps the user to make a rough calculation for the fold difference expression of protein between the samples. To make the accurate calculation for fold difference, user must do the statistical analysis of the data.

- Click on the statistical analysis table for the statistical information of the spots. The data can be used to analyze fold difference between the spots, to determine increased and decreased spot intensity, to generate a histogram for distribution of spots and calculate t-test, ANOVA, and other statistical analysis.

Spot picking: Once the significant spots of interest are identified from the 2D-analysis software, spots can be picked either manually or by robotic arm for protein identification. Spot excision can be carried out on normal difference gel electrophoresis (DIGE)-gels or stained gels soon after the electrophoretic run (Figure 3.17).

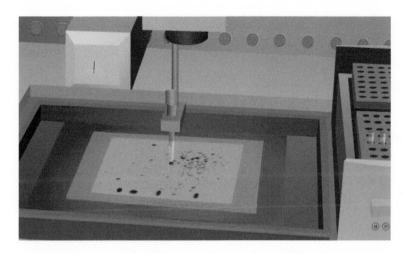

FIGURE 3.17 Methodology for the spot picking.

- Clean the laminar airflow thoroughly with ethanol. Blower is kept ON to avoid any contamination like keratin from user actions. For manual picking, the gel needs to be placed on the supportive material, so that the stained spots are easily visible to naked eyes. User can start manual spot picking with help of scalpel and transfer the gel pieces to tubes. If spot is small, perform manual picking once. If it is large, perform picking twice or thrice to cover the spot completely.
- Transfer the gel pieces into fresh Eppendorf tube with distilled water to avoid shrinking due to evaporation and store at 4°C, till further analysis.
- Before automatic spot picking, the preparative gel from the 2D run electrophoresis needs to be scanned for protein spots to be visible. The preparative gel soon after 2D run needs to be taken for scanning. Depending on landmarks, the robotic arm fixes the co-ordinates for other spots to be picked.
- The DIGE scanning for preparative gel needs to be carried out depending on the settings based on dye used.
- Soon after the instrument is ON, it begins calibration. Users must wait and after calibration, the display pops up for scanning.
- The image from DIGE scanner and image from spot picker need to be overlaid. The DIGE image helps user to annotate the spots and spot picker image helps user to fix the landmarks of spots for the robotic arm to perform picking.
- The user must select the significant spots for global or differential protein expression profiling and annotate them accordingly. Depending on the spots selected, the robotic arm starts picking individual spot in one go. Try to annotate at the middle of the spot, to obtain maximum concentration of the protein from a gel piece.

3.4 Fluorescence-Based Difference Gel Electrophoresis

There are various limitations associated with 2-DE due to gel-to-gel variations and manual artifacts, which emerge mainly from inconsistency in sample preparation, and then from the subsequent gel running itself, during the 1st and 2nd dimensions. These limitations eventually lead to a lack of reproducibility. This has necessitated the need for the development of a technique, DIGE (Figure 3.18) which would overcome these limitations and help in solving the purpose of protein studies in a better way (García-Ramírez et al., 2007). DIGE technique will be discussed in detail in the next chapter.

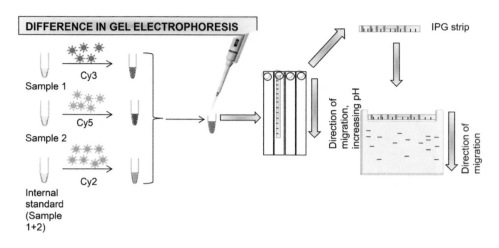

FIGURE 3.18   An overview of the difference gel electrophoresis (DIGE) technique.

3.5 Conclusions

This chapter provides an overview of different gel-based techniques and a brief idea about different applications. The scope of these tools can be further explored depending upon the purpose for which it has to be utilized. Gel-based techniques have been widely used since their inception because of their user-friendliness and the vast amount of data that can be procured from the analysis of these results.

REFERENCES

Brunelle, J. L., & Green, R. (2014). Chapter twelve—One-dimensional SDS-polyacrylamide gel electrophoresis (1D SDS-PAGE). In J. Lorsch (Ed.), *Methods in Enzymology* (Vol. 541, pp. 151–159). Academic Press. https://doi.org/10.1016/B978-0-12-420119-4.00012-4

Gallagher, S. R. (2012). One-dimensional SDS gel electrophoresis of proteins. *Current Protocols in Molecular Biology*, *97*(1), 10.2A.1–10.2A.44. https://doi.org/10.1002/0471142727.mb1002as97

García-Ramírez, M., Canals, F., Hernández, C., Colomé, N., Ferrer, C., Carrasco, E., García-Arumí, J., & Simó, R. (2007). Proteomic analysis of human vitreous fluid by fluorescence-based difference gel electrophoresis (DIGE): A new strategy for identifying potential candidates in the pathogenesis of proliferative diabetic retinopathy. *Diabetologia*, *50*(6), 1294–1303. https://doi.org/10.1007/s00125-007-0627-y

Gygi, S. P., Corthals, G. L., Zhang, Y., Rochon, Y., & Aebersold, R. (2000). Evaluation of two-dimensional gel electrophoresis-based proteome analysis technology. *Proceedings of the National Academy of Sciences*, *97*(17), 9390–9395. https://doi.org/10.1073/pnas.160270797

Longsworth, L. G., & MacInnes, D. A. (1939). Electrophoresis of proteins by the Tiselius method. *Chemical Reviews*, *24*(2), 271–287. https://doi.org/10.1021/cr60078a006

Nijtmans, L. G. J., Henderson, N. S., & Holt, I. J. (2002). Blue Native electrophoresis to study mitochondrial and other protein complexes. *Methods*, *26*(4), 327–334. https://doi.org/10.1016/S1046-2023(02)00038-5

Rabilloud, T. (2002). Two-dimensional gel electrophoresis in proteomics: Old, old fashioned, but it still climbs up the mountains. *PROTEOMICS*, *2*(1), 3–10. https://doi.org/10.1002/1615-9861(200201)2:1<3::AID-PROT3>3.0.CO;2-R

Exercises 3.1

1. In the second dimension of 2-DE, the separation of proteins occurs on which basis?
   a. Charge
   b. Mass
   c. Shape
   d. None of the above

2. In 2-DE, the focused IPG strips are first equilibrated before they undergo the second dimensional separation. What is the role of IAA in equilibration?
   a. Denaturation
   b. Reduction
   c. Alkylation
   d. Oxidation

3. Which of the following is not a component of the cocktail that is prepared for casting SDS gel?
   a. TEMED
   b. DTT
   c. APS
   d. Distilled water

4. A protein showed a molecular mass of 400 kDa when measured by gel filtration chromatography. When this protein was subjected to polyacrylamide gel electrophoresis in the presence of SDS, the protein showed 3 bands with molecular mass of 180, 160, and 60 kDa. When electrophoresis was performed in presence of SDS and dithiothreitol (DTT)/β-mercaptoethanol, again 3 bands were formed but this time with molecular masses of 160, 90, and 60 kDa. What is the subunit composition of the protein?
   a. 60, 90, 90, and 160 kDa
   b. 60, 80, 100, and 160 kDa
   c. 60, 160, and 180 kDa
   d. 60, 85, 95, and 180 kDa

5. Your protein sample has a lot of phosphorylated proteins. In this light, which of the following stain would you prefer to use for the detection of phosphorylated proteins specifically?
   a. Pro-Q Diamond
   b. Coomassie Brilliant Blue
   c. Silver stain
   d. SYPRO Ruby

6. Which of the following statements is *not* correct about isoelectric focusing (IEF)?
   a. Proteins are resolved on the basis of their pI
   b. Involves the use of immobilized pH gradient
   c. Presence of salts interfere the IEF process
   d. Separates proteins on the basis of their size

7. Which one of the following is *not* a feature of Blue Native PAGE?
    a. Provides integrated view of protein function
    b. Identifies multi-protein complexes
    c. Denatures the proteins
    d. Mobility of multi-protein complexes depends on negative charge of bound Coomassie dye

8. Which of the following is TRUE for 2-DE rehydration process?
    a. Salts are removed in the process
    b. Ethanol is added to prevent strip from drying
    c. IPG strip is swelled for protein adsorption
    d. Cysteine residues are alkylated

9. Match the following in columns I and II

Column-I	Column-II
i. Coomassie blue	A. Easily visualized, non-hazardous
ii. Biosafe Coomassie	B. Linear over 3 orders of magnitude
iii. Silver stain	C. High sensitivity
iv. Silver stain plus	D. Most commonly used
v. SYPRO Ruby	E. MS compatibility is an issue, high sensitivity

    a. i-D, ii-A, iii-E, iv-C, v-B
    b. i-D, ii-B, iii-E, iv-A, v-C
    c. i-A, ii-D, iii-E, iv-C, v-B
    d. i-D, ii-E, iii-C, iv-A, v-B

10. Which of the following statement(s) holds true for SDS-PAGE?
    a. Proteins are denatured by the detergent SDS and boiling of sample
    b. Upon SDS treatment, all proteins possess the same charge-to-mass ratio
    c. Smaller proteins migrate more rapidly through the gel
    d. All of the above

11. Your lab mate has expressed some recombinant protein and purified it and you want to know its molecular weight. The information what you have is the gel image. Based on this image you have to guess which kind of experiment he might have performed and what is the molecular weight of the expressed protein?

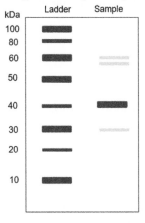

Answers

1. b
2. c
3. b
4. a
5. a
6. d
7. c
8. c
9. a
10. d
11. To check the molecular weight of the purified proteins, one can perform SDS-PAGE experiment. In this experiment a protein ladder (first lane from left) of known molecular weight is used. The molecular weight of the corresponding bands is annotated. From this gel image it can be concluded that expressed protein has molecular weight around 40 kDa. There are some contaminating proteins and/or degradation products also present.

4

2D Difference Gel Electrophoresis (2D-DIGE)

Preamble

Proteome studies carried out by the traditional two-dimensional gel electrophoresis (2-DE) technique have limitations with respect to reproducibility that can be attributed to gel-to-gel variations. These variations affect the quantitative comparison of protein expression levels. To address these issues and enhance the reproducibility, advanced gel-based technique, namely difference gel electrophoresis was reported by (Ünlü et al., 1997). This technique exploits the fact that two different protein samples, when labeled with different fluorescent dyes (Cy3 and Cy5), can be visualized individually on one single gel. These fluorescent dyes are pH insensitive, photostable and spectrally distinct. Moreover, the use of internal control, which is a pool of control and test sample, is labeled with a third fluorescent dye (Cy2) to facilitate co-detection, normalization and accurate quantification of protein samples.

Terminology

- **DIGE:** Difference gel electrophoresis
- **Cy dyes:** Cyanine dyes are used for labeling protein samples from different sources, which can be mixed and run together using the electrophoresis process.
- **Internal control:** A pool of equal amount of the samples under study, which is labeled with a third fluorescent dye (usually Cy2).
- **Co-detection:** Simultaneous detection of protein spots from two different samples, enabled due to the multiplexing nature of DIGE.
- **DeCyder™:** DIGE Image Analysis software.
- **DIA:** Differential In-gel Analysis module.
- **BVA:** Biological Variation Analysis module.

4.1 Workflow for 2D-DIGE

The pooled internal standard for DIGE is prepared by mixing equal amounts of all samples that are being run in the experiment. This overcomes the problems of gel-to-gel variations. Each protein sample as well as the internal standard is labeled with a differently fluorescing cyanine dye (Cy2, Cy3, and Cy5), which allows all protein samples to be simultaneously run on a single gel. The dye binds covalently to the ε-amine group of lysine residues in proteins, the labeled protein samples are mixed and run on a single two-dimensional gel electrophoresis (2-DE) gel. Separation takes place on the basis of isoelectric points (pIs) of the proteins in one dimension and based on molecular weight of the proteins in the second dimension with the smaller proteins migrating further along the gel. The gel containing all the protein samples can be viewed by illuminating it alternately with excitation wavelengths corresponding to the various cyanine dyes (Figure 4.1). Information of molecular weight and pI of proteins, as well as differential expression, can be obtained from these spots.

DOI: 10.1201/9781003098645-6

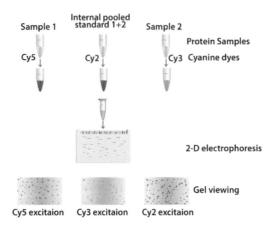

FIGURE 4.1 Difference gel electrophoresis (DIGE).

A typical DIGE experiment involves the following steps:

- Sample preparation and labeling
- IPG strip rehydration
- First dimension (IEF)
- Second dimension (SDS-PAGE)
- Scanning
- Image analysis and data interpretation

4.1.1 Sample Preparation and Labeling

The protein pellet dissolved in appropriate buffer is taken as the starting sample for labeling. The sample pH should be 8.5 and can be adjusted by using 100 mM NaOH. Each of the samples is labeled with one of the fluorescent dyes (e.g., the control sample can be labeled with Cy3 and the test sample can be labeled with Cy5 or vice versa). An internal pool containing an equal amount of the control and treated samples is mixed together and is labeled with a different fluorescent dye, usually Cy2. The dye swapping should be done to check the reproducibility and efficiency of the working protocol. The reaction is quenched using 10 mM lysine, which combines with unbound dye molecules and stops the reaction. The labeled samples are then stored at 4°C until they are further rehydrated.

Cyanine dyes:

- The Cy dyes have different absorbance maxima and emission maxima (Figure 4.2).
- N-hydroxy succinimidyl ester derivatives of cyanine.

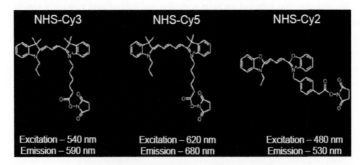

FIGURE 4.2 The excitation and emission wavelengths of cyanine dyes.

- Binds to ε-amine groups of the protein's lysine residues (Figure 4.3).
- Lysine is targeted for cyanine labeling.
- Spectrally resolvable fluors that are matched for mass and charge.
- There is no change in signal over the wide pH range used during first-dimension (isoelectric focusing [IEF]) separation.
- The discrete signal from each fluor with minimal cross talk contributes to high accuracy.

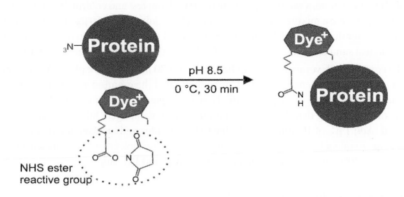

FIGURE 4.3 DIGE: The labeling chemistry.

- Cyanine labeling enables accurate analysis of differential expression of proteins between the samples. It is possible to label three different samples within the same 2D gel, enabling accurate analysis of differences in protein abundance between samples by preventing gel-to-gel variation. Cyanine dyes are used for the differential labeling of the samples. Cy dyes have been designed to be both mass- and charge-matched. Spectrally resolvable with different excitation and emission wavelength (Figure 4.4).

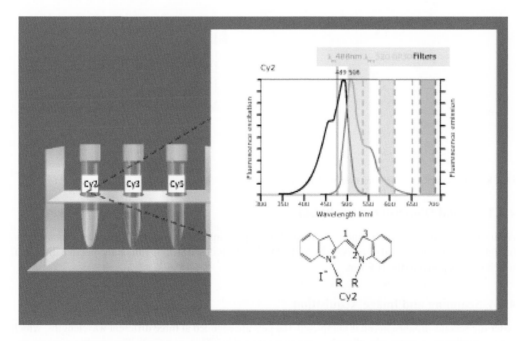

FIGURE 4.4 Spectrally resolvable Cy dyes labeled samples with different excitation and emission wavelength.

A representative protocol for Cy labeling is demonstrated as following.

- Prepare stock solution and working solution as per the desired concentration of the dye. N,N-dimethylformamide (DMF), used to reconstitute the fluors should be of high quality (anhydrous, >99.8% pure). It must not become contaminated with water that can degrade the DMF to amine compounds. The DMF stock solution should be replaced at least every three months. Vortex the content well. The final concentration of dye should be 400 pmol/μL.

- For labeling, take protein sample equivalent to 50 μg from test and control. For internal standard for a single gel, take a mixture of test and control (1:1) equivalent to 50 μg. The pH of the sample should be 8.8 for the accurate labeling of the dye. Add 1 μL each of cy2 to internal standard sample, Cy3 to test samples and Cy5 to control. Cy2 must always be labeled to the internal standard.

- Mix the content well, vortex it, and incubate in ice for 30 min in a dark place. Avoid exposure to light, as dyes are light sensitive. The cyanine dyes label the e-amino group of the lysine in the protein. When a CyDye is coupled to the lysine, it replaces the lysine's single positive charge with its own, ensuring that the pI of the protein does not change. Only 3% of the total protein is labeled. Add 1 μL of 10 mmol/L of lysine to the samples and incubate for 10 min in ice. Free lysine helps to quench the reaction as it binds to free dye molecules. The labeled samples can be stored at −20°C or one can proceed for rehydration and IEF steps.

4.1.2 Rehydration of IPG Strips and IEF

The IPG strips need to be rehydrated with the labeled sample, which is mixed with the respective IPG buffer. The sample is then placed over the IPG strip and rehydration is allowed to occur by incubating the strips at room temperature for 16 hrs. After 1 hr, the samples are overlaid with mineral oil to prevent sample evaporation. The rehydrated strips are further used for IEF (Figure 4.5). For IEF, the rehydrated strips are first placed in the IEF tray and the program for IEF run is selected as per the strip length and pH range. The applied voltage gradient brings about the separation of proteins based on their pI where the net charge on a given protein is zero.

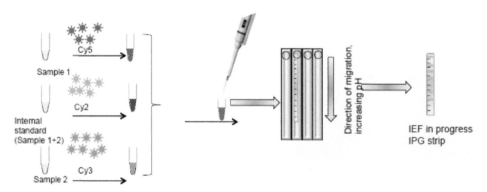

FIGURE 4.5   Sample labeling using cyanine dyes followed by IEF.

4.1.3 Second Dimension (SDS-PAGE)

As discussed in the previous chapter, the separation of proteins on the basis of their molecular weight is brought about in the second dimension by SDS-PAGE. The focused strip is run on the gel in a direction, which is orthogonal to the IEF run.

4.1.4 Scanning and Image Acquisition

After the second-dimension run is completed, the gels are scanned at three different wavelengths, which are the excitation wavelengths of each of the three dyes (Cy3, Cy5, and Cy2). The resulting images are stored for further analysis.

4.1.5 Image Analysis and Data Interpretation

The saved images are further analyzed by using software especially designed for 2D-DIGE, such as DeCyder. The pair-wise analysis between test and control samples can be performed using Differential in-gel analysis (DIA) module, whereas the analysis between multiple samples belonging to two different groups can be performed using the Biological Variation Analysis (BVA) module of the software. Details of the data analysis will be discussed in the next chapter.

4.2 Applications of 2D-DIGE

The 2D-DIGE technique has been used for several proteomic applications. Few representative studies are described here (Table 4.1). In the next coming chapters, in-depth applications of 2D-DIGE will be discussed.

DIGE for detection of markers for cancer: Human tissue biopsies are obtained, and protein is extracted. The 2D-DIGE is carried out as per the described protocol. The control samples of healthy control are labeled with Cy3, whereas the treated samples (diseased patient for cancers) are labeled with Cy5 or vice versa. An internal pool is labeled with Cy2.

TABLE 4.1

Application of 2D-DIGE Technique

Application	Reference
Proteome analysis of human colorectal cancer tissue using DIGE	(Bai et al., 2010)
Biomarker discovery for esophageal cancer	(Uemura et al., 2009)
Whole proteome analysis of drug treated rat muscle	(Kenyani et al., 2011)
Post-translational modifications	(DeKroon et al., 2012)

4.3 Advantages and Limitations of 2D-DIGE

Advantages

- The amount of protein required is very lesser as compared to that in 2-DE. DIGE is extremely sensitive i.e., <1 fmol of protein can be detected and it can also enable the linear detection over a >10,000-fold protein abundance range.

- Higher reproducibility as two different samples can be analyzed on the same gel. Therefore, differences in protein expression levels are purely attributed to biological variations.

- Accurate quantitative comparison, which can be attributed to the use of internal control, which allows normalization of spot intensities across different gels and increases the accuracy of protein expression differences.

- No post-electrophoretic processing (fixation or destaining) is necessary and thereby there is a reduction in protein loss particularly in the low molecular weight range.

- User bias can be eliminated by the co-detection method.

Limitations

- Only two different samples can be analyzed on a single gel. For more than two samples, a large number of pair-wise comparisons would be required. Mass spectrometry-based techniques such as iTRAQ or TMT could be used for 4- or 8-plexing.

- Spot excision is a problem since spots are not visible to the naked eye and hence requires aids such as Robotic Spot picking. Alternatively, a normal 2-DE gel has to be run for these samples, which is then stained with Coomassie dye and spots of interest are excised.

- Fluorescence detection comes with several inherent problems such as high background, the detection of signals from non-protein sources (e.g., dust and residue on plates), and overlap of signals from different fluorophores.

4.4 Comparison of Gel-Based Techniques

Table 4.2 below describes the comparison of different electrophoresis techniques.

TABLE 4.2

Comparison of Different Gel-Based Techniques

Technique	Advantages	Disadvantages	Applications
1-DE	• Easy technique to check for protein purity in a given protein extraction.	• Large number of proteins cannot be separated from a complex mixture with good resolution. • It cannot be used to study whole proteome or to analyze complex fluids like serum or cell lysates.	• Test protein purity. • Study protein expression.
2-DE	• Powerful technique for simultaneous separation of thousands of proteins. • Highly sensitive visualization of proteins as small differences in protein expression levels can be detected with statistical confidence. • Relatively easy to handle and affordable.	• Laborious and time consuming. • Requires 800–1000 µg of protein as the starting material. • In one gel, only one sample can be analyzed. • Requires procedures like staining or fixation to be done after the second-dimension gel electrophoresis	• Study global and differential protein expression and resolution of complex proteins (Chen et al., 2004). • For biomarker discovery (Lescuyer et al., 2007).
DIGE	• Requires less amount of starting protein. • Highly reproducible and sensitive as two different samples can be analyzed on the same gel and so the differences in protein expression levels are purely attributed to biological variations. • Better quantitative comparison is attributed to the use of internal control. • No need for fixation or destaining as fluorescent dyes are used for sample labeling.	• Only two different types of samples can be analyzed on a single gel. • Spot excision is a problem since it is not visible to the naked eye and hence requires aids such as robotic spot picking. • Fluorescence detection gives high background at times when signals from different labels may become mixed.	• In cancer studies (Bai et al., 2010). • For biomarker discovery (Uemura et al., 2009). • For studying post-translational modifications (DeKroon et al., 2012).

4.5 Conclusions

2D-DIGE technique has gained wide acceptance in proteomic studies because of its sensitive quantification abilities and reduced analytical variability. It also offers an advantage over 2-DE with respect to the reproducibility and accuracy in quantification. DIGE is an increasingly employed proteomic technique, which can be used to address several biological questions.

REFERENCES

Bai, X., Li, S.-Y., Yu, B., An, P., Cai, H.-Y., & Du, J.-F. (2010). Proteome analysis of human colorectal cancer tissue using 2-D DIGE and tandem mass spectrometry for identification of disease-related proteins. *African Journal of Biotechnology*, 9(41), 6840–6847. https://doi.org/10.4314/ajb.v9i41

Chen, J.-H., Chang, Y.-W., Yao, C.-W., Chiueh, T.-S., Huang, S.-C., Chien, K.-Y., Chen, A., Chang, F.-Y., Wong, C.-H., & Chen, Y.-J. (2004). Plasma proteome of severe acute respiratory syndrome analyzed by two-dimensional gel electrophoresis and mass spectrometry. *Proceedings of the National Academy of Sciences, 101*(49), 17039–17044. https://doi.org/10.1073/pnas.0407992101

DeKroon, R. M., Robinette, J. B., Osorio, C., Jeong, J. S. Y., Hamlett, E., Mocanu, M., & Alzate, O. (2012). Analysis of protein posttranslational modifications using DIGE-based proteomics. In R. Cramer & R. Westermeier (Eds.), *Difference Gel Electrophoresis (DIGE): Methods and Protocols* (pp. 129–143). Humana Press. https://doi.org/10.1007/978-1-61779-573-2_9

Kenyani, J., Medina-Aunon, J. A., Martinez-Bartolomé, S., Albar, J.-P., Wastling, J. M., & Jones, A. R. (2011). A DIGE study on the effects of salbutamol on the rat muscle proteome—An exemplar of best practice for data sharing in proteomics. *BMC Research Notes, 4*(1), 86. https://doi.org/10.1186/1756-0500-4-86

Lescuyer, P., Hochstrasser, D., & Rabilloud, T. (2007). How Shall We Use the Proteomics Toolbox for Biomarker Discovery? *Journal of Proteome Research, 6*(9), 3371–3376. https://doi.org/10.1021/pr0702060

Uemura, N., Nakanishi, Y., Kato, H., Saito, S., Nagino, M., Hirohashi, S., & Kondo, T. (2009). Transglutaminase 3 as a prognostic biomarker in esophageal cancer revealed by proteomics. *International Journal of Cancer, 124*(9), 2106–2115. https://doi.org/10.1002/ijc.24194

Ünlü, M., Morgan, M. E., & Minden, J. S. (1997). Difference gel electrophoresis. A single gel method for detecting changes in protein extracts. *ELECTROPHORESIS, 18*(11), 2071–2077. https://doi.org/10.1002/elps.1150181133

Exercises 4.1

1. Which of the following is *not* a property of cyanine dyes?
   a. They are spectrally distinct
   b. They are pH insensitive
   c. They are photostable
   d. They become inactive at high pH

2. In Cy dye labeling, what reagent is used to quench the labeling reaction?
   a. 10 mM Lysine
   b. 100 mM Leucine
   c. 10 mM Leucine
   d. 100 mM Lysine

3. DIGE cannot be used for which of the following applications?
   a. To determine post-translational modifications like phosphorylation
   b. To identify differential expression of proteins in brain tissue of healthy control and meningioma patients
   c. To analyze a peptide of molecular weight less than 15 kDa
   d. Whole proteome analysis of a given bacterial strain

4. Gel A has 20 spots and Gel B has 18 spots, of which 10 are common to both gels. The total number of spots in the overlaid image would have?
   a. 38
   b. 18
   c. 30
   d. 28

5. β-Mercaoptoethanol is responsible for breaking which type of interactions in proteins?
   a. Disulfide bond
   b. Hydrophobic interactions
   c. Hydrogen bond
   d. Peptide bond

6. Which of the following should be used to minimize any bias in sample processing in DIGE experiments?
   a. Dye swapping
   b. Addition of cysteine
   c. Use of Cy5 dye
   d. Minimum labeling

7. In case of minimum labeling, cyanine dyes bind to which of the following amino acid?
   a. Arginine
   b. Lysine
   c. Cysteine
   d. Serine

8. In a gel, there are two proteins with different pIs but they have same molecular weight. What will be the position of these protein spots on the gel with respect to each other?
   a. At same position
   b. They will be overlapped
   c. Adjacent to each other
   d. Vertical to each other

9. In DIGE experiments, which of the following factors ensure that each protein on the gel has a reference point?
   a. Use of internal standard
   b. Use of Cy3 dye
   c. High voltage during IEF
   d. None of the above

10. Mahesh's experiment is focused on finding the differentially expressed proteins by doing a comparative study between healthy and disease samples using 2-DE. The result shows a lot of gel-to-gel variation and he might have to repeat the experiment again. What would be your suggestion to Mahesh to improve his results?
    a. He must run more 2D gels to improve the quality
    b. He must check for the artifacts present in the gel
    c. He must properly define the spot boundaries while doing the gel analysis
    d. He must perform a 2D-DIGE gel

11. You have performed a DIGE experiment with aim of identifying differentially expressed proteins in glioblastoma patients. In DIGE, one uses three kinds of dyes, Cy2, Cy3, and Cy5 and after scanning the DIGE gel one obtains three images out of the same gel. In this experiment control sample was labeled with Cy5 and glioblastoma sample was labeled with Cy3. Your task is to identify what the following images represent.

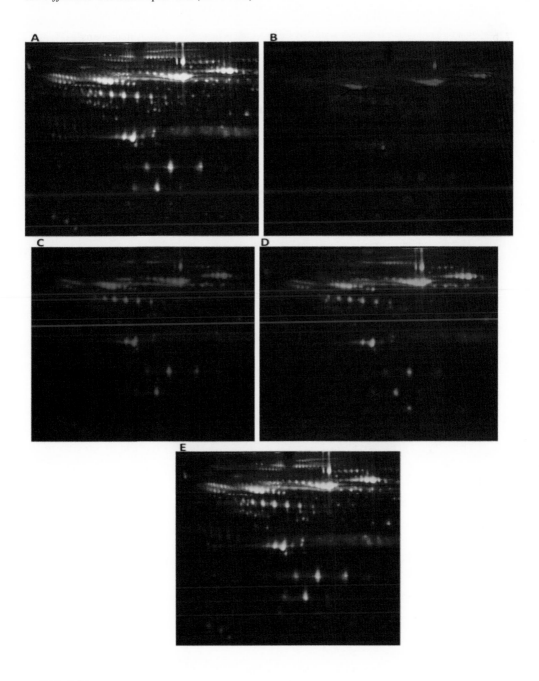

Answers

1. d
2. a
3. a
4. d
5. a
6. a

7. b

8. c

9. a

10. d

11. A: Combined image of control, glioblastoma and internal standard

    B: Internal standard labeled with Cy2

    C: Healthy controls labeled with Cy5

    D: Gliobalstoma samples labeled with Cy3

    E: Overlapping image of C & D

5

Gel-Based Proteomic Data Analysis

Preamble

The aim of gel-based proteomic experiments is the identification of protein spots which are differentially expressed or unique under a given condition (Chevalier, 2010). After the gel is obtained using two-dimensional gel electrophoresis (2-DE) or DIGE, some variations may be observed. These variations might be due to various reasons such as making of the gel, running of the gel, manual handling difference, scanning the gel, and differences or bias due to manual errors in the analysis of the gel images. These differences may be biological or technical. Biological variations may be due to the different types of experimental sets conducted and technical variations may be due to the issues in running the gel or other experimental artifacts. Variations due to any kind may result in erroneous results. For the identification of statistically significant spots, analysis of the obtained gels is very important. The gels are scanned and images are analyzed using image processing software.

5.1 Artifacts in 2D Gels

The gel-based proteomics is mainly utilised for the identification of unique or differentially expressed proteins between the samples (Chevalier, 2010). There are many types of artifacts that may occur in gels and result in bad gels, which cannot be used for analysis (Dunbar, 1987). Some of the commonly occurring artifacts in 2D gels are as follows.

5.1.1 Poor Sample Preparation

At times, even though the gel run is smooth, artifacts may appear in gels (Figure 5.1). The cause for this might be poor sample preparation, which includes:

- Poor solubilization of protein in the buffer.
- Impurities due to running of crude protein sample.
- Improper washing during protein extraction steps.
- Presence of salts in protein sample which interferes with IEF.
- Highly abundant proteins like IgG, in samples can mask the lower abundant proteins.

5.1.2 Horizontal Streaking

Horizontal streaking can occur as a result of different types of erroneous procedures (Figure 5.2). Some of the reasons for streaking on gels are as follows:

- Poor sample preparation.
- Presence of interfering agents such as carbohydrates or nucleic acid.
- Presence of salts.
- Due to incomplete resolution of the spots on gel.

DOI: 10.1201/9781003098645-7

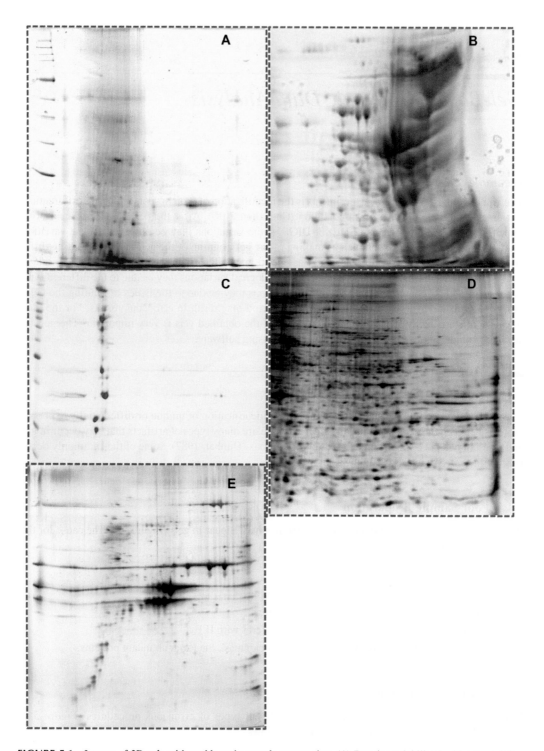

FIGURE 5.1   Images of 2D gels with problems in sample preparation. (A) Proteins solubilized without precipitation, (B) crude sample, (C) improper washing in TCA-acetone extraction, (D) salt impurities, and (E) high-abundant proteins.

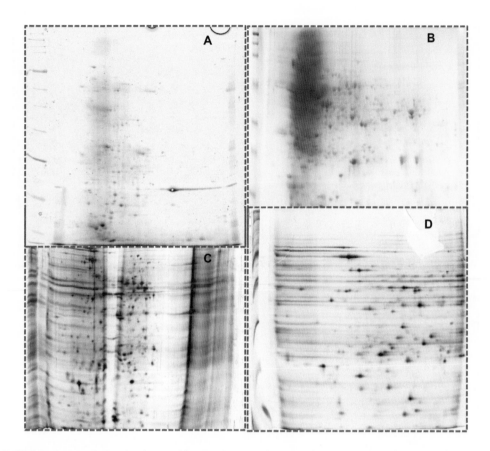

FIGURE 5.2 Images of 2D gels with streaking. (A) Poor sample preparation, (B) presence of impurities like carbohydrates, (C) high salt content, and (D) under-focusing.

5.1.3 Chemical Impurities

Many artifacts arise in gels while preparation of the gel. This may occur due to the poor quality of the chemicals used for the preparation of the gels (Figure 5.3).

- Impurities in acrylamide, bisacrylamide.
- Usage of old N,N,N′,N′-tetramethylethylenediamine (TEMED) – as very less amount of TEMED is required per gel, thus the same stock may be used for a long time.
- Presence of impurities in salts like Tris.
- Impurities in urea – carbamylation train in gels.

5.1.4 Improper Gel Running

At times, if there is no discrepancy with the protein sample preparation or with the chemicals used to prepare the gels, issues may arise due to improper running of the gels (Figure 5.4). For any good 2D gel, proper IEF and the second-dimension electrophoresis are required. Some artifacts on gels may arise due to:

- **Improper isoelectric focusing:** it is important to have the correct program for focusing on the sample – this may lead to under-focusing or over-focusing, which may reflect on the quality of the final gel.
- **Incorrect equilibration step:** either with DTT or with IAA, which results in artifacts on the final gel.
- Improper gel running in the second dimension.

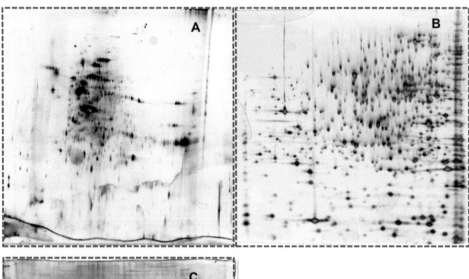

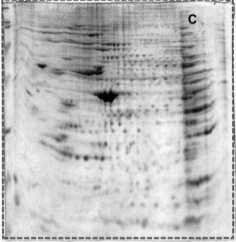

FIGURE 5.3   Images of 2D gels with problems due to chemical impurities. (A) Old TEMED, (B) bad quality tris, and (C) impure urea.

5.2 Scanning of the Gels

For further analysis of the gels, it is necessary to capture an image of the gel spots. For capturing the gel image, the following steps should be considered.

- The gel is laid carefully on the gel image scanner.
- Since the size of the gels is large, care has to be taken while transferring it from the destaining solution to the scanner platform. The gel may break during this process, which results in a black-colored line across the image, thus reducing the gel image quality. This discrepancy may affect the spots detected and affect the gel image analysis.
- Another precaution that needs to be taken before capturing the image is that the gel has to be laid on the platform such that there are no air bubbles between the gel and the surface. This leads to poor image quality. This issue can be resolved by rolling a wet glass rod (dipped in destaining solution) and removing any air bubbles.
- The gel is then scanned using the software. The image is previewed and the colors are adjusted. The image is in grayscale and can be adjusted for the optimum intensity of all the spots.

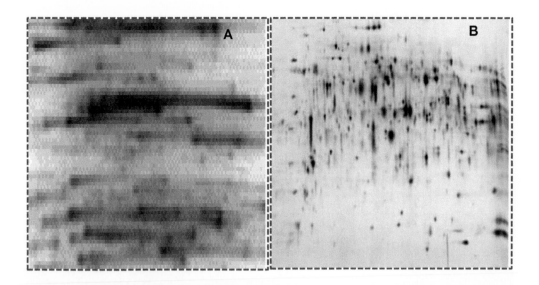

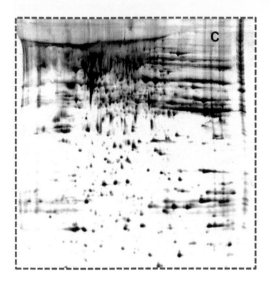

FIGURE 5.4 Images of 2D gels with improper gel run. (A) Under-focusing, (B) inadequate equilibration, and (C) improper second dimension.

5.3 Methods for Analysis of a 2D Gel

After capturing the image on the scanner, image analysis is done for interpretation of the results of the experiment. The ideal method is to use software for automated analysis along with manual validation of results (Figure 5.5). Thus, the results shown by the software must be verified by the user to confirm that spots are real and not artifacts. There are many software available to analyze 2D gels. Table 5.1 provides a list of a few software.

Following two-dimensional gel electrophoresis (2-DE), the gels are scanned using a suitable scanner, and the images obtained are analyzed by means of various available software. These allow interpretation of data present in the form of spots on the gel images. Following are few points related to the image analysis using software.

TABLE 5.1

Software Used for Gel Image Analysis of Traditional 2D Gels

2-DE Software	Website
ImageMaster 2D platinum	http://gelifesciences.com
PDQuest	http://discover.biorad.com (Kaynar, 2012)
Delta 2D	http://www.decodon.com
Dymension	http://www.syngene.com
Ludesi 2D gel image analysis	http://www.ludesi.com
Progenesis	http://www.nonlinear.com

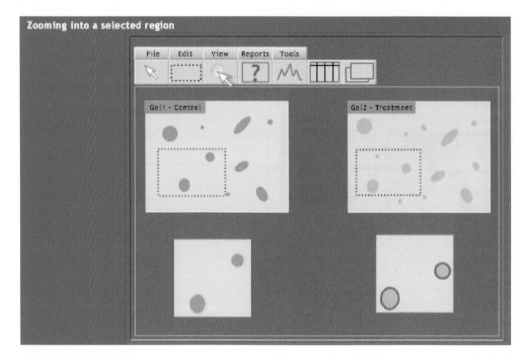

FIGURE 5.5 A snapshot for software-based 2D gel analysis.

- It is possible to load either a single or multiple gel images simultaneously. This is done by means of the "load" option in the file menu. The saved gel images must be chosen, which are then displayed on the software. Several tools are available for analysis of the gels. It is possible to crop the gels by selecting a specific region that is to be studied and then selecting the "crop gels" function. Cropping gels help in selection of regions with high spot density or in trimming of regions with high background staining but no spots.

- Specific selected regions of the gel can be zoomed for viewing the spots more closely and for comparison of spots between two gels. This is particularly useful for gels having large number of spots. Overlaying of images is a particularly useful tool for comparison of two gels. The gel images are overlaid such that they appear merged and spots that coincide will overlap with each other. This is extremely helpful while comparing clinical samples as control and diseased/treated, providing clear indication of the proteins that are differentially expressed.

- The spots on the gels can be displayed as a 3-dimensional (3D) graph. Either the entire gel can be chosen or a particular region can be selected for this representation. The peaks obtained in the graphical representation are directly related to the spot intensity. Every spot on the gel can be detected by selecting the "detect spots" option. Parameters such as "smoothness", "saliency", and "min area" should be adjusted for maximum clarity. Once this is done, each spot will either be encircled or marked with a cross, depending on the settings, along with the spot numbers.
- The software facilitates interpretation of the gel images by matching two different gel images obtained. The matching spots are marked and any variations in the spot position are indicated by blue lines in one of the gels. This provides an understanding about the reproducibility across gels. Information regarding the various physical parameters of each spot can be obtained via the spot table. This table provides the spot number, intensity, area, volume and saliency of the spots. These parameters help in judging the quality of the gel.
- In addition to physical parameters, several statistical parameters can also be computed for each spot on the gel such as central tendency, mean, median, dispersion, coefficient of variation, standard deviation, etc. Scatter plots and histograms can also be plotted for clear data analysis. These provide information regarding intra- and inter-gel variations. It is also possible to specifically compare a particular selected spot across gels. When the gel is run with molecular weight markers, the weight of unknown proteins can be estimated from these. It is also possible to estimate the isoelectric point of the proteins. These parameters, in addition to other physical and statistical parameters, can be obtained for each spot.
- Software for DIGE analysis varies with respect to certain features compared to normal 2-DE analysis. It can compare three gels simultaneously, of which one is typically the pooled internal standard containing all spots. Any changes implemented in one gel such as cropping, spot selection, etc., will be implemented across all three gels in DIGE. Other features and tools for DIGE analysis are same as those used for 2-DE analysis. Physical and statistical parameters of all spots on the gels can be determined through their corresponding reports. Three-dimensional graphical representation provides information regarding the intensity of spots on the gel.

There are specific protocols to be followed depending on the software used. ImageMaster 2D Platinum (IMP7) software has a user-friendly interface (Figure 5.6) (Ananthi et al., 2011). The protocol followed for image analysis using IMP7 software is given below.

STAGE 1:

1. The scanned images which are saved in the .mel or .tiff format are first imported to the image pool of the IMP7 image analysis software from the source folder.
2. It is important to keep in mind that all the images which are to be analyzed are of the same dimensions, contrast and brightness conditions. This should be ensured before importing the images.
3. Within the IMP7 software, a new project is created with the required label.
4. A new workspace is then created in the project and all the images (control as well as treated; for all the reps under study) are considered in the workspace for analysis.
5. Now a match-set, which comprises all the gel images (control as well as treated; for all the reps under study), is created.
6. This is followed by spot detection. The spots can also be detected automatically by the software or manually. In manual detection of spots, the spot boundaries are individually marked in each of the gel images. The contrast and zooming features can be adjusted for better resolution of spot boundaries.

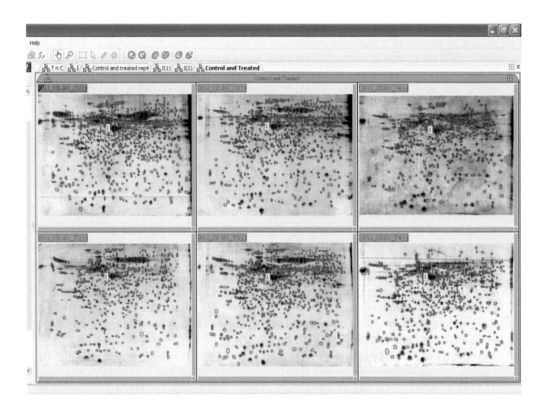

FIGURE 5.6   Representation of the IMP7 software interface for setting landmarks.

STAGE 2:

1. Once all the spots are manually detected in each of the gels, the next step of matching gels can be carried out (Figure 5.7). Ensure that the detected spots are approximately equal in number in each gel.

2. Create match classes and then include gel images in the respective classes of "Control" and "Treated".

3. Display images belonging to one of the 2 classes. For instance, take up "Control" class. It has the gel images with spots that were manually detected. The first gel is taken as the "Reference gel" when the individual gels are being matched amongst themselves.

4. Alternatively, all the gels can be considered in a single match class and then be compared. The first gel will be taken as "Reference gel", by default. This can be changed by simply dragging the required gel in the first position in the order of the created class.

5. Define landmarks (i.e., spots that are prominent and common to each of the gels that are being compared). Ensure that the entire gel area is covered when you are defining landmarks. Optimally 4–5 landmarks are sufficient for analysis.

6. The gels are now ready to be matched within the said class.

7. Click on "Match gels" icon on the toolbar. It will now display the matched gels and the number of total matches that are identified. The match count includes the spots, which are present in all the gels under consideration.

8. From the Match Analysis Table, the ANOVA values for each of the matches can be obtained. The matches with ANOVA value <0.05 are considered to be significant. These spots are then checked if they are matched properly. The software may miss out pre-marked spots where manually checking of the 3D views of the spots becomes essential.

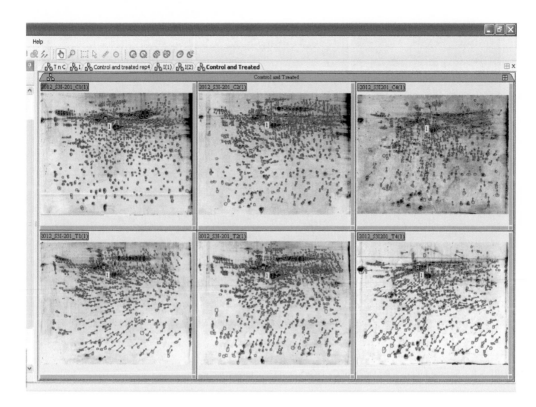

FIGURE 5.7 Matching of gels in IMP7 software.

9. Subsequently, if the spots are missed out in any of the gels, they could be compared and marked. Again, statistical analysis has to be performed.
10. Calculate the ratio which is average of treated intensities/average of control intensities. Correspondingly, note down the trends of spots, i.e., whether down-regulated or up-regulated.

5.4 DIGE Analysis Using DeCyder

The traditional 2D gels are stained using dyes like Coomassie brilliant blue. However, the 2D-DIGE utilizes labeling of proteins using the fluorescent cyanine dyes. There are three types of cyanine dyes – Cy3, Cy5, and Cy2 – which are used for labeling the control, test, and internal control samples. For this, control, test, and internal control samples are run in the same gel. Different images can be obtained from the same gel corresponding to each type of sample (Figure 5.8). Images of the gel are captured using scanner and 2D-DIGE gel analysis can be performed using the DeCyder software (Karp et al., 2004).

Steps for using the DeCyder software for analysis of gels are as follows.

Step 1: Image Loading

The first step in DIGE analysis is loading the relevant images in the software. For this, the image loader option is selected. The .tiff and .GEL files obtained from the Typhoon scanner are used for acquiring the images in the software. All 3 images – Cy2, Cy3, and Cy5 – for one gel are selected and loaded (Figure 5.9). This image loading can be done for as many biological replicates as needed for the analysis.

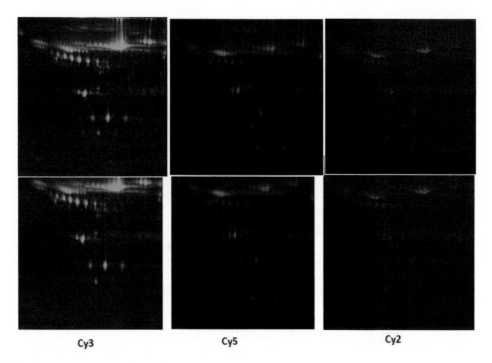

FIGURE 5.8   Scanned images for DIGE of Cy3, Cy5, and Cy2, respectively.

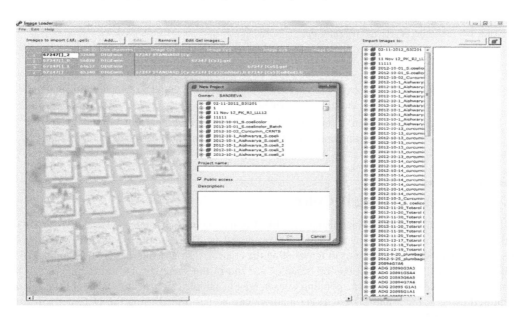

FIGURE 5.9   Representation of image loader in DeCyder.

Step 2: Image Editing

All the imported gel images are selected and the "Edit Gel Images" tab is selected. This view shows the Cy2 images i.e., standards of all the gels. The Cy2 images can be cropped and rotated as required in this window. The changes made to Cy2 gels are applied to the Cy3 and Cy5 gels as well i.e., other gels are cropped using the same dimensions. All the changes are saved and the window is closed.

Step 3: Image Import

The new project icon at the right corner of the screen is clicked, and new file is created and named. The cropped images are imported into this folder.

Step 4: Image Analysis

It can be done using differential in-gel analysis (DIA), biological variation analysis (BVA), and extended data analysis (EDA).

Differential In-gel Analysis (DIA)

For analysis of the Cy2, Cy3, and Cy5, images of one set DIA are used. For this, a new DIA workspace is created and relevant images are imported into the workspace. Exclusion filters can be set to exclude certain areas because there is no distinction between real protein spots and dust particles (Figure 5.10). The estimated numbers of spots are added and the images are processed. After processing, the DIA provides a 3D view of all the spots and the max slope and max volume for all the spots (Figure 5.11). The max slope for a real protein spot is between 1 and 1.5.

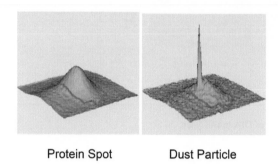

Protein Spot   Dust Particle

FIGURE 5.10   Image of protein spot and dust particle in DIGE using DeCyder.

Biological Variation Analysis (BVA)

The BVA is used for analyzing the Cy2, Cy3, and Cy5 gels of more than one biological replicate, e.g., to compare samples from different patients. The batch processor also helps in comparison of a large number of gels simultaneously. For this, all the gels to be analyzed are selected. The DIA and BVA batch-list are created to proceed further (Figure 5.12).

After this, one of the Cy2 gel is selected as the master gel, and other gels are matched against this gel. A new BVA workspace is created. Two groups are made as control and test, and the gels are sorted appropriately. BVA has different modes as below.

- **Spot Map Mode:** Only the standard gels are shown in this mode
- **Match Mode:** All the gels can be matched in this mode. It provides an option for adding new matches, creating new spots, merging spots, and splitting the spots, which are demarcated by the software. Then, a match-set can be created which incorporates all the changes. This process helps in the accurate matching of the gels; however, verification should be done manually. This tool is especially helpful for marking those spots, which are present but have not been detected by the software.
- **Protein Mode:** The protein mode shows the statistical values for all the gels. It also shows the number of gels in which that spot has appeared. The values of the selected protein parameters: Student's T-test, one-way ANOVA, two-way ANOVA, and the fold change are displayed for each and every spot, along with the spot number on the master gel (Figure 5.13).

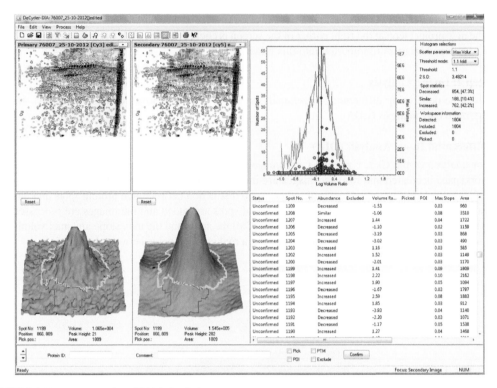

FIGURE 5.11   Representation of DIA in DeCyder.

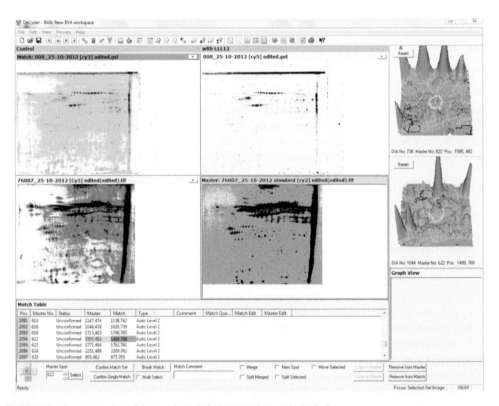

FIGURE 5.12   Representation of the match mode in BVA workspace in DeCyder.

Pos.	Master ...	Status	Protein...	Appearan... ▽	T-test	Av. Ratio	1-ANO...
1	576	Unconfirmed		9 (9)	0.0020	-2.47	0.0020
2	1153	Unconfirmed		9 (9)	0.0023	1.42	0.0023
3	1169	Unconfirmed		9 (9)	0.0098	1.87	0.0098
4	536	Unconfirmed		9 (9)	0.011	3.73	0.011
5	424	Unconfirmed		9 (9)	0.012	-1.42	0.012
6	813	Unconfirmed		9 (9)	0.018	4.01	0.018
7	642	Unconfirmed		9 (9)	0.026	2.39	0.026
8	486	Unconfirmed		9 (9)	0.027	-2.08	0.027
9	859	Unconfirmed		9 (9)	0.027	1.93	0.027
10	1038	Unconfirmed		9 (9)	0.027	-1.59	0.027
11	887	Unconfirmed		9 (9)	0.033	1.97	0.033
12	1293	Unconfirmed		9 (9)	0.033	-2.31	0.033
13	1266	Unconfirmed		9 (9)	0.039	-2.38	0.039
14	566	Unconfirmed		9 (9)	0.041	1.55	0.041
15	515	Unconfirmed		9 (9)	0.045	-2.12	0.045
16	1488	Unconfirmed		9 (9)	0.052	1.55	0.052
17	326	Unconfirmed		9 (9)	0.053	-1.92	0.053
18	488	Unconfirmed		9 (9)	0.062	-2.38	0.062
19	659	Unconfirmed		9 (9)	0.063	1.75	0.063
20	647	Unconfirmed		9 (9)	0.064	4.10	0.064

FIGURE 5.13　Table for protein mode in DeCyder software.

- **Appearance Mode:** Appearance mode displays a comparison of peak heights, etc., for every component.
- **3D View:** The 3D view for each spot can be seen and compared with one another (Figure 5.14). This tool is especially helpful to compare the spots chosen by the software and to detect spots where improper matching has been done. This mode is also useful to detect the authenticity of a spot – whether it is a protein spot or a dust particle.

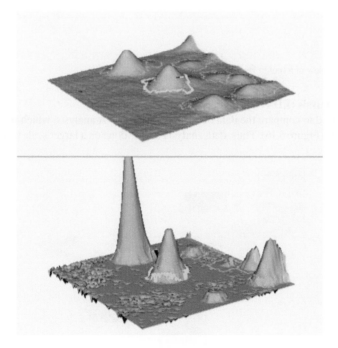

FIGURE 5.14　Representation of 3D view of a spot in DeCyder software.

- **Graph View:** The graph view shows the relative abundance of the protein in each gel (Figure 5.15). This is done for individual spots. The graph view plots the values of the control samples and their average as well as for the experimental group. The graph view also enables to know whether a particular spot is up-regulated or down-regulated.

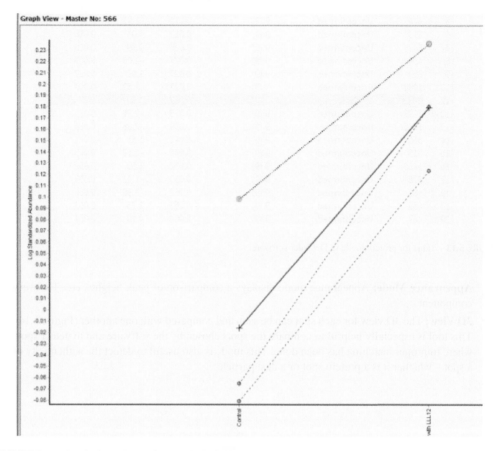

FIGURE 5.15   Graph view of a spot in DeCyder software.

Extended Data Analysis (EDA)

The EDA can be used to compare the data obtained from the BVA analysis, which includes all the individual DIA analysis (Figure 5.16). Thus, data analysis can be done on a larger scale using EDA. EDA can

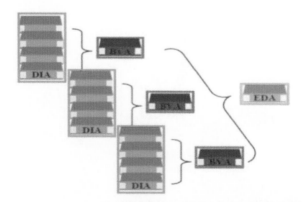

FIGURE 5.16   Representation for EDA. (Figure modified from – DIGE applications and data analysis, GE manual.)

be used to compare different datasets. DIA can be used to find out if any groups or clusters are present in the data. We can also find whether there are proteins which follow a pattern or behave in a certain way. This may help to identify protein spots, which behave similarly as other spots across gels, and may give information for the trends of those proteins.

EDA may be applied to four types of analysis.

a. **Principal components analysis (PCA):** The principal component analysis is used to find whether there are outliers in the data (Figure 5.17). It is also used to find whether protein expression is uniform in multiple samples from the same experimental group.

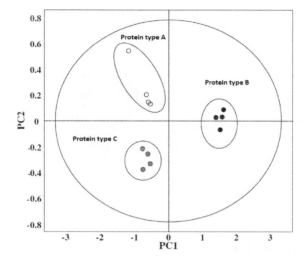

FIGURE 5.17   PCA analysis based on protein clusters.

b. **Differential expression analysis:** This is done using statistical tests like Student's T-test or one-way ANOVA derived from the DIA data.
c. **Pattern analysis:** Pattern analysis enables clustering of the data to find patterns in the expression profiles in EDA data (Figure 5.18). For this, no prior information about the variables

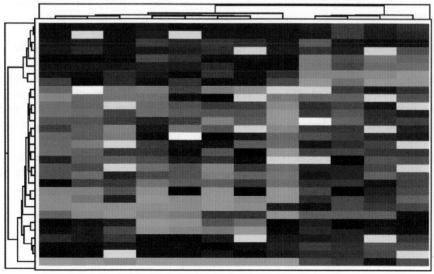

Each column reprsents the Cy3 and Cy5 labeled proteins of interest

FIGURE 5.18   Pattern analysis in EDA for hierarchical clustering.

used is required. Items with similar expression profiles become clustered like proteins, spot maps, and experimental groups. The results from pattern analysis are represented as a hierarchical cluster displayed as a heat map with dendrogram, showing different classes in the data set.

d. **Discriminant analysis:** The discriminant analysis is used for classifying the unknown data into known classes (Figure 5.19). This helps in the identification of the proteins that may be indicators of a particular disease, i.e., this analysis may help in the identification of biomarkers for diseases.

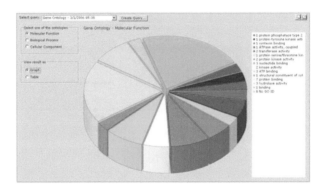

FIGURE 5.19   EDA – discriminant analysis.

Representation of Data Using EDA

An example of dataset consisting of data from two different infectious diseases analyzed using EDA is given in Figure 5.20a and b. The data shows a comparative analysis of two diseases as compared to control.

Scanning of 2D-DIGE Gel

After running 2D-DIGE, the scanner uses the property of dye used for labeling the sample to produce a protein profile image. The dye when exposed to a different wavelength, due to excitation and emission property of the dyes they fluoresce and subsequently the images can be captured using the scanner (Figure 5.21).

- When the instrument is turned ON, it will go for automatic calibration to check all the connections. Take out the cassette for scanning. Cassette provides the platform for gels to scan. A hollow cassette is used to scan large format gels in the glass plates. For small gels, cassette with glass platform is used. The cassette needs to be placed inside the scanner. Proper fit is necessary for better scanning. The user needs to set the parameters for scanning and specify the scanning area depending on the size of the gel. For gel format, predefined gel format can be selected or user can define the gel format using user selection option.
- Selection of the channel check boxes must be performed in sequential numerical order. The excitation and emission wavelength help to produce optimum results (Zhang et al., 1998).
  - Cy2: Ex/Em: 480/520 for 0.35 exposure.
  - Cy3: Ex/Em: 540/595 for 0.90 exposure.
  - Cy5: Ex/Em: 635/680 for 0.70 exposure.
- User can set the exposure time depending upon sample quantity used.
- Once all the settings are done, user can start scanning. Before scanning, the software asks to save the image. Provide the location under the folder and name the gel properly.

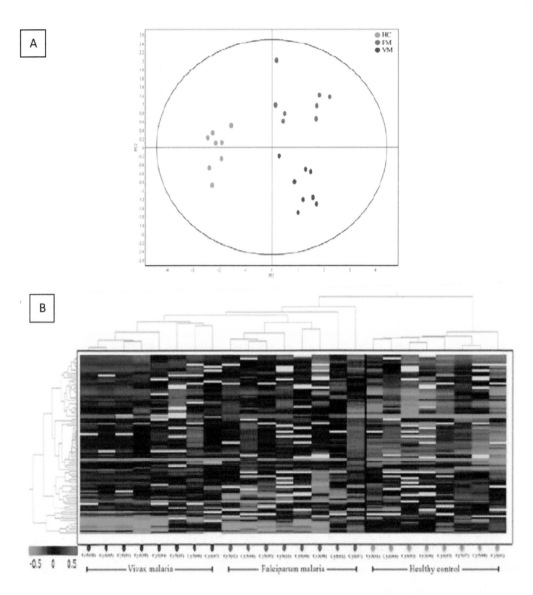

FIGURE 5.20A Data analyzed from EDA. The EDA analysis was conducted for control (healthy subject) against two tests – 2 different types of infectious diseases. (A) PCA analysis – clustering of different phenotypical classes. The PC1 component separates the control group from the other set, while the PC2 clusters the diseased groups separately. (B) Dendrogram showing the separation of different experimental groups after the hierarchical cluster analysis (red – represents up-regulated; green – represents down-regulated; black – represents spots with no significant change in expression level).

- During scanning, the image is displayed row by row for each channel. In case the image shows some red spots in the protein region, stop the scan and re-scan with less exposure time. The display provides the scanning report.
- Cropping and image editing depend on user requirements and just for appearance purpose user can make the changes. Cropping helps to remove unwanted area from analysis.
- In case the image gel plate is kept on "-" side, while scanning the gels can be kept either within the glass plate or it can be performed after removing it from the plates. Software provides option for user to rotate the image and later on save option. Now the gel image is ready to be uploaded and analyzed by software tools (Figure 5.22).

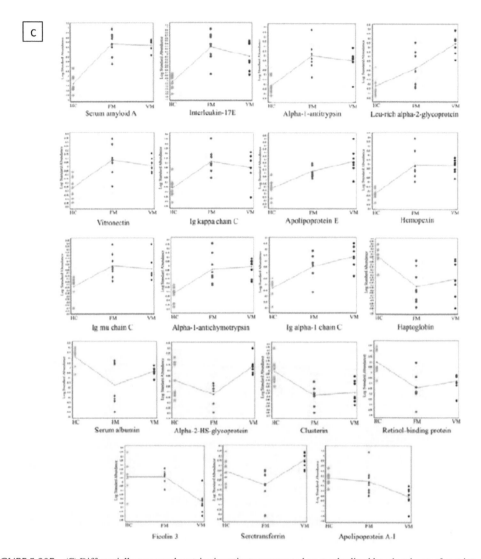

FIGURE 5.20B   (C) Differentially expressed proteins in patients represented as standardized log abundance of spot intensity.

FIGURE 5.21   2D-DIGE gel scanning.

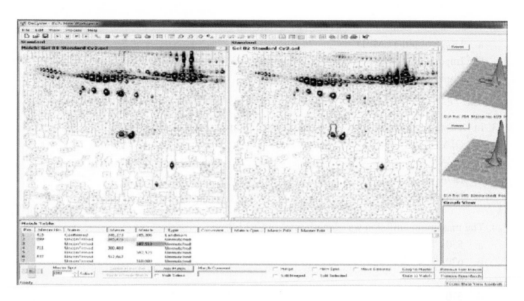

FIGURE 5.22 2D-DIGE gel analysis. After running 2D-DIGE, the protein profile obtained needs to be analyzed for the protein spot of interest, showing significant fold change and appearing in most of the samples. As demonstrated, the DeCyder software helps to minimize system variation and enables the identification of significant spots of biological importance.

REFERENCES

Ananthi, S., Santhosh, R. S., Nila, M. V., Prajna, N. V., Lalitha, P., & Dharmalingam, K. (2011). Comparative proteomics of human male and female tears by two-dimensional electrophoresis. *Experimental Eye Research*, 92(6), 454–463. https://doi.org/10.1016/j.exer.2011.03.002

Chevalier, F. (2010). Highlights on the capacities of "Gel-based" proteomics. *Proteome Science*, 8(1), 23. https://doi.org/10.1186/1477-5956-8-23

Dunbar, B. S. (Ed.). (1987). Troubleshooting and artifacts in two-dimensional polyacrylamide gel electrophoresis. *Two-Dimensional Electrophoresis and Immunological Techniques* (pp. 173–195). Springer US. https://doi.org/10.1007/978-1-4613-1957-3_10

Karp, N. A., Kreil, D. P., & Lilley, K. S. (2004). Determining a significant change in protein expression with DeCyder™ during a pair-wise comparison using two-dimensional difference gel electrophoresis. *PROTEOMICS*, 4(5), 1421–1432. https://doi.org/10.1002/pmic.200300681

Kaynar, O. (2012). PDQuest: Precise & reliable tool for 2-DE gel analysis. *Lucrări Ştiinţifice – Medicină Veterinară, Universitatea de Ştiinţe Agricole Şi Medicină Veterinară "Ion Ionescu de La Brad". Iaşi*, 55(1/2). https://www.cabdirect.org/cabdirect/abstract/20123340069

Zhang, T., Chen, C., Gong, Q., Yan, W., Wang, S., Yang, H., Jian, H., & Xu, G. (1998). Time-resolved excited state dynamics of a cyanine dye. *Chemical Physics Letters*, 298(1), 236–240. https://doi.org/10.1016/S0009-2614(98)01165-8

https://www.gelifesciences.com/gehcls_images/GELS/Related%20Content/Files/1314742967685/litdoc375-635_typhoon_userguide_20110831023220.pdf

http://www.med.unc.edu/sysprot/files/DeCyder%20Manual%20from%20GE.pdf

Exercises 5.1

1. What can lead to vertical streaking in gels?
   a. Second dimension run at high voltage
   b. Usage of incorrect pH strip for IEF
   c. IEF settings
   d. Impurities in protein sample

2. What leads to formation of carbamylation trains in gels?
   a. Usage of old TEMED
   b. Impurities on tris
   c. Impurities in urea
   d. Impurities in acrylamide

3. In DeCyder, which function would be used for comparing large datasets?
   a. DIA
   b. BVA
   c. EDA
   d. None of the above

4. The maximum emission wavelength (nm) of Cy3 is?
   a. 550
   b. 670
   c. 650
   d. 570

5. Which factor should not be taken into account when comparing the spots on different gels?
   a. Smoothness
   b. Saliency
   c. Minimum area
   d. Dispersion

6. In 2-DE gel analysis software, which parameter is used to compare gels on a spot-by-spot basis?
   a. Crop tool
   b. Zoom tool
   c. Image overlaying
   d. Spot analysis

7. If 60 μg of protein sample A is labeled with Cy3 and 60 μg of protein sample B is labeled with Cy5, what should the internal control be composed of?
   a. 30 μg of protein sample A labeled with Cy2
   b. 30 μg of protein sample B labeled with Cy2
   c. 60 μg of standard protein like BSA labeled with Cy2
   d. 30 μg of protein sample A and 30 μg of protein sample B labeled with Cy2

8. Which of the following analyses can be performed in BVA (biological variation analysis) module of DIGE data analysis?
   a. Principal component analysis
   b. More than one technical replicate
   c. More than one biological replicate
   d. None of them

9. Which of the following parameters is used to create an identical spot boundary across all the channels in 2D-DIGE gel analysis?
   a. Detection
   b. Co-detection
   c. BVA
   d. EDA

10. Which of the following components leads to the clogging of an immobilized pH strip (IPG) and may form complexes with proteins by electrostatic interactions?
    a. Polysaccharides
    b. Nucleic acids
    c. Lipids
    d. All of the above

11. You have performed a DIGE experiment with the intention of identifying differentially expressed proteins in breast carcinoma patients and healthy individuals. After scanning the DIGE gel three images are obtained from the same gel. Image analysis is performed and the excerpt of the result is displayed in the following table. Your task is to arrange these spots in descending order based on their significance in this experiment.

Spot Id	Master No.	Appearance	T-Test	Fold-Change	1-ANOVA
1	1341	9 (9)	0.00048	−2.05	0.00048
2	850	9 (9)	0.00031	2.81	0.00031
3	535	9 (9)	0.10092	−5.66	0.10092
4	449	9 (9)	0.00003	1.82	0.00003
5	634	9 (9)	0.0014	18.56	0.0014

Answers

1. D
2. C
3. C
4. D
5. D
6. D
7. D
8. C
9. B
10. D
11. In this experiment, to find the most significant proteins spots first criteria is to look for T-test and ANOVA values. Since only one spot is statistically insignificant, the next parameter one can look for is the fold-changes. The order is represented in the table.

Spot ID	Master No.	Appearance	T-Test	Fold Change	1-ANOVA
5	634	9 (9)	0.0014	18.56	0.0014
2	850	9 (9)	0.00031	2.81	0.00031
1	1341	9 (9)	0.00048	−2.05	0.00048
4	449	9 (9)	0.00003	1.82	0.00003
3	535	9 (9)	0.10092	−5.66	0.10092

Module III

Basics of Mass Spectrometry and Quantitative Proteomics

6

Introduction to Mass Spectrometry

Preamble

Mass spectrometry (MS) is the most versatile technique in the emerging field of proteomics (Ong et al., 2003). MS is a technique for the production of charged molecular species and their separation by magnetic and electric fields based on mass to charge ratio. Unlike genomics, where the availability of resources as whole-genome sequence and microarray chips have revolutionized the growth of the field, proteomics appears to be a challenging field when it comes to identifying proteins, post-translational modifications, or differential expression of proteins subject to different conditions. MS has emerged as an excellent tool capable of performing several applications. MS has emerged from an analytical technique to identify molecules based on the mass, to an advanced form capable of identifying complex molecules such as proteins and their features like post-translational modifications and sequence. Various advancements in soft ionization techniques and combinations of mass analyzers are used to achieve these objectives (Yates et al., 2009). The inclusion of high-performance liquid chromatography (HPLC) with MS has increased the resolution and sensitivity of the technique in which peptides are separated and identified by the mass spectrometer.

Terminology

- **MS:** Mass spectrometry is the technique for protein identification and analysis by the production of charged molecular species in a vacuum and their separation by magnetic and electric fields based on mass to charge (m/z) ratio.
- **RP:** Reverse-phase chromatography is a technique used to separate analytes based on their hydrophobicity.
- **SCX:** Strong cation exchanger is a technique used to separate analytes based on their positive charges.
- **MALDI:** Matrix-assisted laser desorption ionization (MALDI) is an efficient process for generating gas-phase ions of peptides and proteins for mass spectrometric detection. The target plate with the dried matrix-protein sample is exposed to short, intense pulses from a UV laser.
- **ESI:** It is known as electrospray ionization. Ions are formed by spraying a dilute solution of analyte (sample) at atmospheric pressure from the tip of a fine metal capillary, creating a fine mist of droplets. The droplets are formed in a very high electric field and become highly charged. As the solvent evaporates, the peptide and protein molecules in the droplet pick up one or more protons from the solvent to form charged ions.
- **TOF:** Time of flight is a mass analyzer in which the flight time of the ion from the source to the detector is correlated to the m/z of the ion. An ion trap makes use of a combination of electric and magnetic fields that captures ions in a region of a vacuum system or tube. It traps ions using electrical fields and measures the mass by selectively ejecting them to a detector. This analyzer typically has a lower resolution.
- **Quadrupole:** Quadrupole is a type of mass analyzer that uses oscillating electrical fields to selectively stabilize or destabilize the paths of ions passing through a radio frequency (RF) quadrupole field.

DOI: 10.1201/9781003098645-9

6.1 Principle of Mass Spectrometry

The first step in Mass spectrometry (MS) involves generating charged particles in the gaseous form, which can be accelerated using a strong electric field and then analyzed by the detector (Pitt, 2009).

$$M + e^- \rightarrow M^{*+} + 2e^-$$

In the case of proteomic analysis, the proteins are digested with a suitable protease (usually trypsin) and the peptide fragments so obtained are ionized to generate charged particles. These charged particles are accelerated in a vacuum under the presence of an external electric field. They are further fragmented (in tandem MS/MS before finally being detected by the detector. The data is obtained in the form of relative abundance vs. m/z ratio. Relative abundance refers to the abundance of that particular ion in the sample. Using the m/z ratio, an idea about the fragmentation pattern and hence the empirical formula of the molecule or peptide sequence of proteins can be established.

6.2 Components of the Liquid Chromatography Mass Spectrometer

The mass spectrometer is an instrument that produces charged molecular species in a vacuum, separates them by means of electric and magnetic fields, and measures the mass-to-charge ratios and relative abundances of the ions thus produced. It is being increasingly used for the detection and analysis of proteins from complex samples.

A mass spectrometer has three basic components – ionization source, mass analyzer, and detector (Figure 6.1).

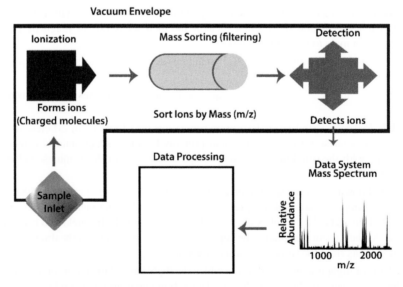

FIGURE 6.1 The basic components of the mass spectrometer.

The mass spectrometer can be coupled with an HPLC system to facilitate easy separation of peptides, prior to their identification. The columns utilize the inherent properties of the peptides to separate them in solution, thereby increasing the resolution of the mass spectrometer. Proteins are digested using trypsin and pre-fractionated using HPLC. MS device makes use of a combination of the ion source and one or two mass analyzers (Figure 6.2).

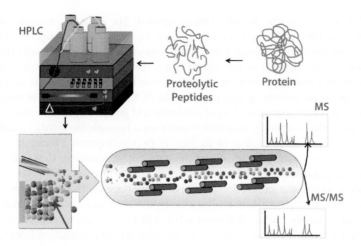

FIGURE 6.2 Schematic of proteomic experiments using HPLC tandem MS.

6.2.1 Chromatographic Unit

MS is a versatile tool for protein identification and quantification. However, it is subjected to various limitations since the detection is solely based on the m/z ratio. It might happen that two different peptide fragments are generated, having same m/z ratio where assigning their true identity becomes a problem. Also, sometimes the highly abundant peptides overshadow the signals of low abundant peptides which might be clinically relevant from the point of view of biomarker discovery. Therefore, fractionation of complex mixtures of peptides into smaller fractions, which can be analyzed accurately and with high sensitivity, is required. The separation methodology using various types of HPLC techniques like reverse-phase liquid chromatography (RPLC), strong cation exchanger (SCX), affinity chromatography and hydrophilic interaction liquid chromatography (HILIC) is an effective pre-fractionation strategy. RPLC and SCX are more commonly used to separate peptides.

RPLC: The columns for RPLC are highly hydrophobic C-18 columns, which strongly interact with the hydrophobic patches of the peptides. The major advantage of RPLC is that the mobile phase used to elute peptides is compatible with ESI. With varying degrees of hydrophobicity, the peptides elute at different time intervals and are detected by MS. A typical RP-HPLC chromatography setup consists of the solvent bottles, degassifier, dual or quaternary pump, sample injector, column, and detector. Different solvents can be placed in the solvent bottles depending upon the purification requirement. These solvents are mixed in the desired ratio and pumped into the column during elution after removal of any trapped air inside it by means of the degassifier (Figure 6.3).

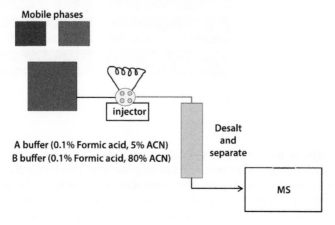

FIGURE 6.3 Schematic representation for RP-HPLC chromatography.

The sample injector system may be automatic or manual. The automatic sampler uses a syringe to inject the sample placed in a vial directly into the column. Once the sample is injected, the mobile phase flows into the column through the pump. The column consists of a stationary matrix that preferentially binds certain analytes. The outlet from the column enters the flow cell where it is detected.

Various stationary phase matrices are available that separate the components of the mixture based on different principles. One of the commonly used matrices, the SCX, separates charged peptides based on their electrostatic interactions with negatively charged sulphonic acid groups on the resin surface. Reverse-phase chromatography is another commonly used tool, which uses a hydrophobic matrix consisting of long aliphatic carbon chains. These retain analytes on the basis of their hydrophobic interactions and can be eluted by changing the polarity of the solvent. Nano-liquid chromatography, which uses C-18 capillary columns, has gained popularity for proteomic studies due to its ability to achieve fine separation. The separated components pass from the column outlet into the flow cell present in the detector. The most commonly used detector for protein analysis is UV detector, which analyzes the protein absorbance at 280 nm and plots a graph of retention time against intensity. Each peak corresponds to a particular analyte in the sample mixture.

SCX: The columns for SCX consist of strong cation exchangers like aliphatic sulfonic acid groups that are negatively charged in an aqueous solution. Tryptic peptides containing positive charge are specifically exchanged using these columns. Using suitable solvents of high ionic strengths, strong ionic interactions are broken. Working with only SCX yields very poor resolution of peptides. However, combining SCX with RPLC yields better resolution, where two different properties of the peptides are utilized in fractionating the peptides.

6.2.2 Ionization Source

The ionization source is responsible for converting analyte molecules into gas-phase ions in a vacuum. This has been made possible by the development of soft ionization techniques, which ensure that the non-volatile protein sample is ionized without completely fragmenting it. An ionization source ionizes the peptides so that they acquire a charge and hence can be accelerated in vacuum under the influence of an external electric field. The ionization source must be effective for all types of molecules (polar, non-polar, non-volatile, etc.) and must be capable of ionizing the analyte without degradation. Various ionization sources are described in Figure 6.4. The most commonly used ionization sources are matrix-assisted laser desorption ionization (MALDI) and electrospray ionization (ESI).

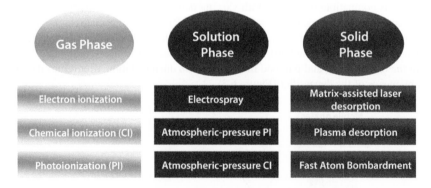

FIGURE 6.4 Different ionization sources: gas phase, solution phase, and solid phase.

6.2.2.1 MALDI

In MALDI, the analyte of interest (e.g., trypsin-digested peptide) is mixed with an aromatic matrix compound like α-cyano-4-hydroxycinnamic acid and sinapinic acid. This is then dissolved in an organic solvent and placed on a metallic sample plate. The evaporation of solvent leaves the analyte embedded in the matrix. The target plate is placed in a vacuum chamber with high voltage and short laser pulses are applied. The matrices absorb the energy from the laser and transmit it to the analyte (peptide fragments), which undergo rapid sublimation resulting in gas-phase ions. These ions then accelerate toward the mass analyzer based on their mass-to-charge ratio (Jurinke et al., 2004). Singly charged ions are produced using MALDI ionization and these ions are accelerated through a vacuum (Figure 6.5).

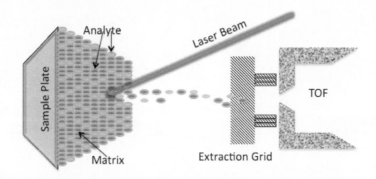

FIGURE 6.5   A schematic diagram of ionization using MALDI.

6.2.2.2 Electrospray Ionization (ESI)

The need for softer ionization techniques, to preserve peptide integrity, led to the emergence of ESI. Trypsin-digested peptides are allowed to pass through a very small capillary, the tip of which is maintained at a very high voltage. As the peptides emerge from the capillary in the form of fine droplets, the droplets become ionized (Figure 6.6). All types of charges are formed in ESI but it is the positively charged ions that are accelerated.

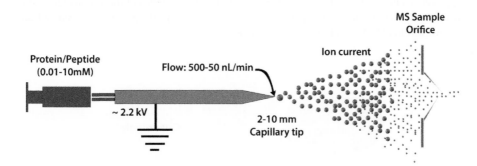

FIGURE 6.6   A schematic diagram of ionization using ESI.

In ESI, the sample is present in the liquid form, and ions are created by spraying a dilute solution of the analyte at atmospheric pressure from the tip of a fine metal capillary, creating a mist of droplets. The droplets are formed in a very high electric field and become highly charged. As the solvent evaporates, the peptide and protein molecules in the droplet pick up one or more protons from the solvent to form charged ions. These ions are then accelerated toward the mass analyzer depending upon their mass and charge.

MALDI and ESI are efficient ionization techniques and their development was awarded the Nobel Prize in 2002. However, each one has several pros and cons, and the type of ionization can be chosen based on the experimental objective (Figure 6.7).

Other ionization sources include atmospheric pressure photo-ionization (APPI) and atmospheric pressure chemical ionization (APCI), where photons generated from specialized UV lamps ionize the peptides in case of APPI and inert gases like charged nitrogen gas transfer their ionic charge to the analyte in case of APCI.

Comparison between MALDI and ESI		
	MALDI	**ESI**
1. Sample analysis	Simple peptide mixture	Analysis of complex sample
2. Bias	Polar/charged peptides	Nonpolar peptides
3. Effect of salts	Salt tolerant	Salt Sensitive
4. Liquid chromatography	Offline	Online, analysis can be coupled to LC
5. Sequence coverage	Less	More
6. Nobel prize	Chemistry, 2002	Chemistry, 2002

FIGURE 6.7   Comparison of MALDI and ESI.

6.2.3 Mass Analyzer

The mass analyzer represents the component of the mass spectrometer where the ionized peptides are separated according to their m/z ratio (Scigelova & Makarov, 2006). Various types of mass analyzers are available (Figure 6.8), some of which are described here. The mass analyzer resolves the ions produced by the ionization source on the basis of their mass-to-charge ratios. Various characteristics such as resolving power, accuracy, mass range and speed determine the efficiency of these analyzers. Commonly known mass analyzers include TOF, quadrupole (Q) and ion trap along with an advanced analyzer; Fourier Transform Ion Cyclotron Resonance (FT-ICR).

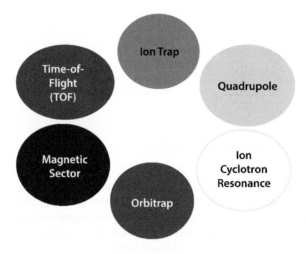

FIGURE 6.8   Different types of mass analyzers used for proteomics applications.

6.2.3.1 TOF

Time of flight (TOF) involves accelerating the charged peptides through a long tube known as a flight tube, maintained at vacuum. The different sized peptides by virtue of their difference in kinetic energy move at different rates and reach the detector. Ions of lower masses are accelerated to higher velocities and reach the detector first. Usually, TOF is employed in association with MALDI and hence, singly charged peptides are accelerated through TOF. The TOF under such circumstances is inversely proportional to the square root of the molecular mass of the ion. Thus, heavier peptides take a longer time to reach the detector than smaller peptides (Figure 6.9). The time required by a peptide to move across the entire flight tube is given by:

$$t = \left(\frac{m}{2qV0} \right)^{\frac{1}{2}} L$$

where

t = time of flight (s)
m = mass of the ion (kg)
q = charge on ion (C)
V_0 = accelerating potential (V)
L = length of flight tube (m)

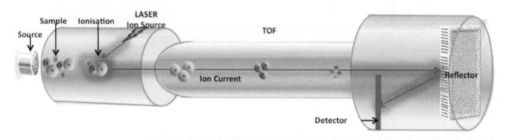

FIGURE 6.9   Illustration of flight (TOF) mass analyzer and its working principle.

6.2.3.2 Quadrupole/Triple Quadrupole

A quadrupole mass analyzer consists of a parallel set of four metallic rods, which are maintained at different potential differences, allowing particular ions to pass through them (Chernushevich et al., 2001). A triple quadrupole (TQ) consists of an additional collision cell in between the two quadrupoles (Figure 6.10) (Hopfgartner et al., 2004). The first quadrupole allows a particular set of ions to move into the collision cell, where that particular ion is fragmented into several ion pieces, all of which are then moved into the third quadrupole and hence the detector.

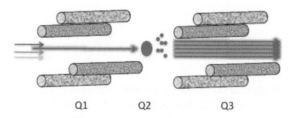

Q1   Q2   Q3

FIGURE 6.10   A typical scheme of a triple quadrupole mass analyzer.

Quadrupole mass analyzers use oscillating electrical fields to selectively stabilize or destabilize the paths of ions passing through a RF quadrupole field. The quadrupole mass analyzer can be operated in either the RF or scanning mode. In the RF mode, ions of all m/z are allowed to pass through which are then detected by the detector. In the scanning mode, the quadrupole analyzer selects ions of a specific m/z value as set by the user. A range can also be entered in which case only those specific ions satisfying the criteria will move toward the detector and the rest are filtered out.

6.2.3.3 Ion Trap

An ion trap mass analyzer consists of two sets of electrodes; the cap electrodes and the ring electrodes maintained at two different voltages. The ions that have a threshold m/z value more than the corresponding m/z value set by the voltage remain trapped in the ring electrode while other ions are allowed to pass into the cap electrode, where they are further fragmented and analyzed (Figure 6.11). An ion trap makes use of a combination of electric and magnetic fields and captures ions in a region of a vacuum system or tube. It traps ions using electrical fields and measures the mass by selectively ejecting them to a detector.

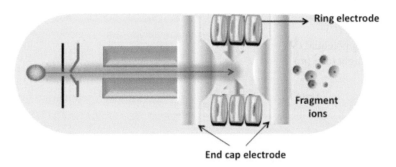

FIGURE 6.11   A typical scheme of ion trap mass analyzer.

6.2.3.4 Fourier Transform Ion Cyclotron Resonance (FT-ICR)

Ion cyclotron resonance is the most advanced form of mass analyzer where the ions of a particular m/z ratio orbit around a pole subjected to a uniform magnetic field. This gives the highest resolution. However, it is extremely complicated to handle (Cho et al., 2015).

6.2.4 Mass Detector

The mass detector is the final component of the mass spectrometer. It records the ion flux or the abundance of ions reaching the surface and hence the m/z ratio. Typically, a detector is an electron multiplier or an ion-to-photon converter (photo multiplier). This is because the amount of ion reaching the detector is extremely small to be detected and hence multiplication is required.

6.3 Tandem MS

One of the most important applications of MS is protein sequencing. The Edman degradation method of protein sequencing is limited to a stretch of approximately 40 amino acids and also with the availability of a free N-terminal. MS-based amino acid sequencing is hence a better option when it comes to protein identification. Tandem MS refers to another round of peptide fragmentation, analysis, and detection after the first round. Technically, it is represented as MS^N, where N takes up an integer value. In tandem MS/MS, one of the peaks obtained is chosen and that particular peptide is allowed to move into the next mass analyzer where it is further fragmented and detected (Figure 6.12). Ideally, the different daughter peptides so obtained differ by one or more amino acids and hence from there, the sequence of the peptide can be determined.

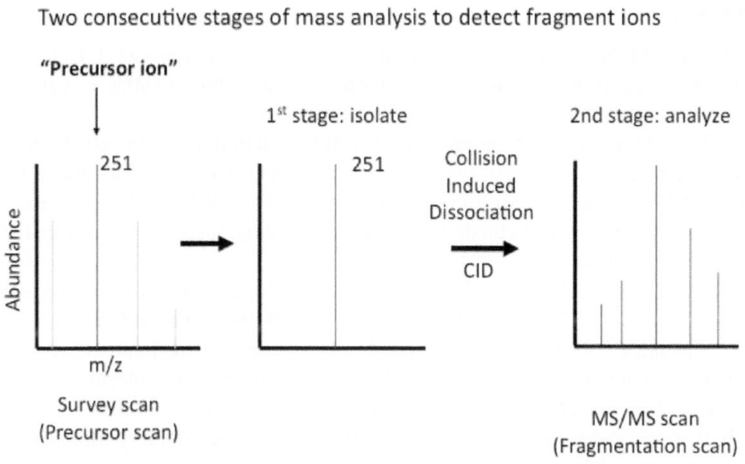

Two consecutive stages of mass analysis to detect fragment ions

"Precursor ion"

1st stage: isolate   2nd stage: analyze

251   251   Collision
                                        Induced
                                        Dissociation

Abundance

CID

m/z

Survey scan
(Precursor scan)

MS/MS scan
(Fragmentation scan)

FIGURE 6.12 Representation of tandem mass spectrometry.

6.4 Applications and Challenges in Clinical Proteomics

All disorders have some signature molecules for diagnosis, and clinical proteomics aims to identify and characterize proteins involved in those disorders. For example, haptoglobin is one of the biomarkers identified in glioblastoma (Grade IV). One of the most challenging aspects of this approach is the abundance of these biomarkers in the samples. The human serum accounts for almost 70% albumin and 25% immunoglobulin G (IgG). The low molecular weight proteins account for less than 2% of human serum. Amidst the cloud of albumin and IgG, all these biomarkers are hidden. Two-dimensional electrophoresis cannot resolve all of these proteins neither can the most sophisticated MS. The low molecular weight proteins are either hidden under the intense signal of the most abundant ions or are so low in concentration that the sensitivity limit of the MS cannot detect it. MS is also very sensitive to levels of contaminations like salt and other biomolecules, which interfere with the fragmentation pattern of the peptides.

These challenges are mitigated by the use of depletion columns, which specifically remove the high abundant proteins and thereby enrich the samples for low-abundant proteins, which could be potential biomarkers (Betzen et al., 2015). Over years, advancements in MS have increased the detection of analytes. Also, the sequential elution of peptides using HPLC columns enhances the sensitivity of the MS. Nonetheless, the technology is still developing and with every passing year, it is turning into a more robust platform for use in advanced clinical diagnosis (Cravatt et al., 2007).

6.5 Conclusions

The success of MS in the area of proteomics has been phenomenal. Many biomarkers have been discovered using MS, which no other proteomic technique could identify. However, like other techniques, MS is also subject to certain limitations, some of which are discussed in this chapter. The main issues with MS are its sensitivity and resolution. With the development of sophisticated technologies, the sensitivity and resolution power of the instrument are increasing, which is extremely advantageous for clinical proteomics applications. The pace of proteomics research has been phenomenal due to the advancement of a strong MS platform. It can be said that MS can survive without proteomics but the converse is not true.

REFERENCES

Betzen, C., Alhamdani, M. S. S., Lueong, S., Schröder, C., Stang, A., & Hoheisel, J. D. (2015). Clinical proteomics: Promises, challenges and limitations of affinity arrays. *PROTEOMICS – Clinical Applications*, *9*(3–4), 342–347. https://doi.org/10.1002/prca.201400156

Chernushevich, I. V., Loboda, A. V., & Thomson, B. A. (2001). An introduction to quadrupole–time-of-flight mass spectrometry. *Journal of Mass Spectrometry*, *36*(8), 849–865. https://doi.org/10.1002/jms.207

Cho, Y., Ahmed, A., Islam, A., & Kim, S. (2015). Developments in FT-ICR MS instrumentation, ionization techniques, and data interpretation methods for petroleomics. *Mass Spectrometry Reviews*, *34*(2), 248–263. https://doi.org/10.1002/mas.21438

Cravatt, B. F., Simon, G. M., & Yates III, J. R. (2007). The biological impact of mass-spectrometry-based proteomics. *Nature*, *450*(7172), 991–1000. https://doi.org/10.1038/nature06525

Hopfgartner, G., Varesio, E., Tschäppät, V., Grivet, C., Bourgogne, E., & Leuthold, L. A. (2004). Triple quadrupole linear ion trap mass spectrometer for the analysis of small molecules and macromolecules. *Journal of Mass Spectrometry*, *39*(8), 845–855. https://doi.org/10.1002/jms.659

Jurinke, C., Oeth, P., & van den Boom, D. (2004). MALDI-TOF mass spectrometry. *Molecular Biotechnology*, *26*(2), 147–163. https://doi.org/10.1385/MB:26:2:147

Ong, S.-E., Foster, L. J., & Mann, M. (2003). Mass spectrometric-based approaches in quantitative proteomics. *Methods*, *29*(2), 124–130. https://doi.org/10.1016/S1046-2023(02)00303-1

Pitt, J. J. (2009). Principles and applications of liquid chromatography-mass spectrometry in clinical biochemistry. *The Clinical Biochemist Reviews*, *30*(1), 19–34.

Scigelova, M., & Makarov, A. (2006). Orbitrap mass analyzer – Overview and applications in proteomics. *PROTEOMICS*, *6*(S2), 16–21. https://doi.org/10.1002/pmic.200600528

Yates, J., Ruse, C., & Nakorchevsky, A. (2009). Proteomics by mass spectrometry: Approaches, advances, and applications. *Annual Review of Biomedical Engineering*, *11*, 49–79. https://doi.org/10.1146/annurev-bioeng-061008-124934

Exercises 6.1

1. The most sensitive mass analyzer is?
   a. TOF
   b. Quadrupole
   c. Ion trap
   d. FT-ICR

2. MALDI produces _____ charged ions.
   a. Single and positive
   b. Single and negative
   c. Multiple and positive
   d. Multiple and negative

3. The most common type of chromatographic technique used prior to MS is?
   a. SXC and RPC
   b. Size exclusion and RPC
   c. RPC
   d. Affinity and RPC

4. Tandem mass spectrometry can be used for which of the following purposes?
   a. Protein sequencing
   b. Biomarker discovery
   c. Metabolite detection
   d. All of the above

5. Which of the following is not an ionization source?
   a. Fast atom bombardment
   b. Matrix-assisted laser desorption ionization
   c. Ion cyclotron resonance
   d. Electrospray ionization

6. From the given options, choose the correct sequence of events which take place during MALDI analysis.
   a. Sample loading, ionization, mass analyzed by TOF, laser bombardment, detection of ions, spectra generation
   b. Sample loading, laser bombardment, ionization, mass analyzed by TOF, detection of ions, spectra generation
   c. Sample loading, mass analyzed by TOF, detection of ions, laser bombardment, ionization, spectra generation
   d. Mass analyzed by TOF, detection of ions, laser bombardment, ionization, sample loading, spectra generation

7. In mass spectrometry, which of the following roles does the mass analyzer play?
   a. Sample introduction
   b. Sample ionization
   c. Ion mass-to-charge filtering
   d. Data acquisition

8. Different ionization sources are categorized into gas, solid, and solution phase. Which of the following ionization sources falls under the category of gas phase ionization?
   a. ESI
   b. MALDI
   c. Electron ionization
   d. All of the above

9. A purified protein sample was subjected to in-solution digestion, followed by mass spectrometry data acquisition using MALDI-TOF/TOF instrument. What should the charge state be in the MS/MS ion search?
   a. +1
   b. +2
   c. +1, +2, +3
   d. None of the above

10. Which one of the following statements is true with regards to triple quadrupole mass spectrometry?

    a. The third quadrupole scans all the ions in the radio frequency mode to generate a spectrum based on the varying behavior of ions in an oscillating electrical field
    b. The first quadrupole scans the ions coming from the ionization source
    c. Consists of two sets of parallel metallic rods interspersed by collision cells
    d. Ions which enter the collision cell are fragmented by collision against an inert gas like argon

11. You have been provided with a crude sample of a nuclear extract from a eukaryotic cell. You wish to identify the level of two transcriptional factors in response to induction by Human Papilloma Virus.

    The two transcription factors are identified to generate signature peptides as:

    TF1: ....Leu-His-Val-Ile-Ala-Gly-Pro-Met-Leu-Pro-Ile-Lys

    TF2: ....Gln-His-Leu-Asn-Ala-His-Val-His-His-Arg-Pro-Ile-Lys

    (1) How were the peptides generated? Would you find any anomaly in the generation of the peptides?
    (2) Suggest the type of column that should be used to significantly resolve these two peptides if they have almost similar size. According to the type of column used, which peptide will be eluted first from the column?
    (3) Suggest the type of ionization source that should be used?

Answers

1. d
2. a
3. a
4. d
5. c
6. b
7. c
8. c
9. a
10. c
11. (1) On close analysis of the peptides it can be inferred that peptide 1 has many non-polar residues while peptide 2 has many polar basic side chains. The C-terminal of both the peptides have lysine, which indicates that the peptides are generated by trypsin digestion of the proteins in question, as trypsin digests right after lysine or arginine residues. However, trypsin fails to digest the peptides if the residue next to lysine or arginine is proline and this is the reason why trypsin fails to digest past arginine residue in TF2 peptide.

    (2) For best possible resolution, the best columns to be used are strong cation exchangers followed by reverse-phase columns. The peptide 2 will elute first.

    (3) The type of ionization source for better ionization should be electrospray ionization because it is a softer ionization technique where the chance of the peptide remaining intact is more.

7

Hybrid Mass Spectrometry Configurations

Preamble

As discussed in the previous chapter, mass spectrometry (MS) is by far the most versatile technique used in proteomics. We had also discussed some of the limitations of MS and how to deal with them, in terms of sample preparation. The two most important features required for a mass spectrometer are sensitivity and resolution. Sensitivity refers to the ability of an instrument to detect minute amounts of the analyte, whereas resolution refers to the ability of a mass spectrometer to resolve different molecular species with similar but distinct masses. Both of these features need to be addressed properly during the usage of MS in clinical proteomics because certain biomarkers are present in the sample in the range of femtomoles and hidden by the large abundant molecules. Further, during tryptic digestion, certain peptides may be generated having extremely close molecular mass. Assigning the correct identity solely based on mass becomes challenging, which can be mitigated by a highly resolvable instrument. All these problems have been addressed by configuring the MS instrumentation. In this chapter, we will discuss innovations in the MS techniques using hybrid MS configuration.

Terminology

- **Mass spectrometry (MS):** Technique for production of charged molecular species and their separation by magnetic and electric fields based on mass to charge ratio.
- **Mass analyzer:** The mass analyzer resolves the ions produced by the ionization source on the basis of their mass-to-charge ratios. Various characteristics such as resolving power, accuracy, mass range, and speed determine the efficiency of these analyzers.
- **Mass resolution:** Ability of a mass spectrometer to resolve different molecular species with similar but distinct masses.
- **Mass accuracy:** It determines how close a mass measurement is to its true (theoretical or exact) value.
- **Top-down:** An analytical approach of separating and analyzing intact proteins. It involves direct analysis of intact proteins, without previous proteolytic digestion.
- **Bottom-up:** An analytical approach of separating and analyzing peptides following proteolytic digestion of the sample.
- **Time of flight (TOF):** A mass analyzer in which the flight time of the ion from the source to the detector is correlated to the m/z of the ion.
- **Ion traps:** An ion trap makes use of a combination of electric and magnetic fields that captures ions in a region of a vacuum system or tube. It traps ions using electrical fields and measures the mass by selectively ejecting them to a detector.
- **Quadrupole:** Quadrupole mass analyzers use oscillating electrical fields to selectively stabilize or destabilize the paths of ions passing through a radio frequency (RF) quadrupole field.
- **Ion cyclotron resonance (ICR):** A high-frequency mass spectrometer in which the specific ions to be detected are selected by setting a value of the quotient mass/charge, after which they absorb maximum energy through the effect of a high-frequency electric field and a constant magnetic field perpendicular to the electric field.

DOI: 10.1201/9781003098645-10

- **Orbitrap:** An orbitrap is a type of MS analyzer that consists of an outer barrel-like electrode and a coaxial inner spindle-like electrode that form an electrostatic field with the quadro-logarithmic potential distribution. Ions that are injected tangentially into the electric field cycle around the central electrode rings oscillate along the central spindle. The ions are detected by means of the frequency of their harmonic oscillations, which in turn is dependent on the m/z ratio.
- **Magnetic sector:** Double-focusing magnetic sector mass spectrometers make use of static electric and magnetic fields for the detection of ions. They provide high sensitivity, high resolution, and reproducibility.
- **Collision cell:** A device that selects a specific ion and further fragments it into smaller ions.
- **Linear triple quadrupole (LTQ):** Two separate quadrupoles placed at two edges of a collision cell.

7.1 Requirement of a Hybrid MS Configuration

Before delving into the various hybrid MS configurations, it is essential to know why there is a requirement for combining different mass analyzers to give hybrid configurations (Glish & Burinsky, 2008). The various mass analyzers discussed in the previous lecture had one or more limitations. For example, the mass resolution of quadrupole was poor while that of TOF was medium, whereas the mass range of ion trap and quadrupole was low. On the other hand, FT-ICR which gave the best resolution and mass coverage was extremely complicated to operate. All these problems surfaced while studying various clinical samples, where the range of analytes may vary up to magnitudes of 10^8 (e.g., human serum) and the discovery of low molecular mass analytes require high sensitivity and mass accuracy. A complex sample will contribute to various types of analytes. In the area of clinical proteomics, many proteins having a wide concentration range may be present, and often these proteins act in regulating cellular pathways. On trypsin digestion, the number of peptides generated may have overlapping properties, leading to their co-elution. Adding to the challenge is the need to detect very low-abundance species in the presence of highly abundant ones.

Tandem MS makes use of a combination of an ion source and two different mass analyzers (hybrid MS configurations), separated by a collision cell, in order to provide improved resolution of the fragment ions (McLafferty, 1981; Sleno & Volmer, 2004). The mass analyzers may either be the same or different. The first mass analyzer usually operates in a scanning mode in order to select only a particular ion, which is further fragmented and resolved in the second analyzer. This can be used for protein sequencing studies. The technology of MS is too versatile but individual mass analyzers have limitations hence scientists came up with the brilliant idea of combining the techniques so that loopholes in one technique could be complemented by the strength of the other. This marked the origin of different combinations of hybrid mass analyzers.

7.2 Various Types of Hybrid MS Configurations

Few hybrid MS configurations were developed to meet the needs of the complex samples (Figure 7.1). Various hybrid-MS configurations will be described in the following section.

   a. **TOF-TOF:** Joining two time-of-flight tubes.
   b. **Triple-quadrupole:** Joining two quadrupoles at the ends with a collision cell in between.
   c. **Q-TOF:** Joining a quadrupole configuration with the time-of-flight tube.
   d. **LTQ-FTICR:** Joining triple quadrupole with Ion cyclotron resonance (ICR).

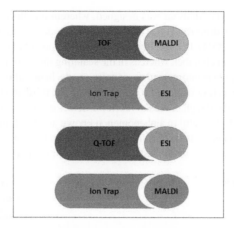

FIGURE 7.1 Typical MS configurations. The ionization source and mass analyzer can be combined in different ways to give varying configurations for the mass spectrometer. Some of the most commonly used MS configurations are MALDI with TOF, ESI with ion trap, ESI with Q and TOF, and MALDI with ion trap.

7.2.1 TOF-TOF

TOF or time-of-flight mass analyzers separate ions by virtue of their kinetic energy. Trypsin-digested peptides with different masses have different kinetic energies. Although the m/z range of TOF was large, the resolution was poor. Combining another TOF tube with the existing tube increased the path length of the ions and hence resulted in better resolution compared to a single TOF. However, the addition of another TOF tube resulted in increased complexity of the instrument. This problem was mitigated using reflectron, by which the same TOF tube could be used twice for mass analysis. The reflectron uses a constant electrostatic field to reflect the ions. Higher energy heavier ions penetrate deep into the reflectron and hence are reflected later and become slower, leading to a higher resolution (Figure 7.2). The same path length now becomes twice the path length and hence the time taken becomes double as it would have been for a single TOF tube (Suckau et al., 2003). An additional collision cell may be introduced in between to facilitate better resolution.

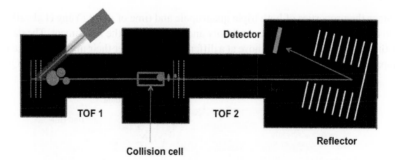

FIGURE 7.2 Tandem MS: MALDI-TOF-TOF-MS. A common tandem MS configuration in which the ions are first resolved on the basis of their time-of-flight in the first TOF analyzer. The selected ions enter the collision cell where they are further fragmented. The fragmented ions are accelerated and further resolved on the basis of their m/z values in the second TOF tube, after which they are detected.

7.2.2 Triple Quadrupole

The linear quadrupole consists of four sets of parallel metallic rods maintained at different potential differences to facilitate sequential elution of ions towards the detector. However, the resolution of quadrupole was very poor as more than two closely related peptides would reach the detector simultaneously. To address the problem, a triple quadrupole mass analyzer was designed, wherein a collision cell separated two quadru-poles (Figure 7.3). The triple quadrupole (TQ) is operated in two modes i.e., radio frequency mode and scan-ning mode (Hopfgartner et al., 2004). In radio frequency mode, the potential difference is set so that all ions are allowed to move into the next chamber, whereas in the scanning mode, only selective ions are allowed to move. In TQ, the first quadrupole is maintained at scanning mode, hence allows selective ions to move into the collision cell where the ion is fragmented into daughter ions. The third quadrupole is maintained at radio frequency mode, allowing all the daughter ions to reach the detector and thus have an idea about the fate of the parent ion. This is particularly useful in a phenomenon known as MRM, multiple reactions monitoring, where the fate of the parent ion is studied. This finds applications in both pharmaceuticals and proteomics, where a target analyte (e.g., protein) is monitored through a specific method prepared in MRM mode.

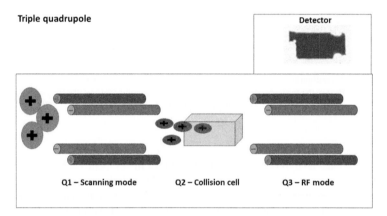

FIGURE 7.3   Tandem MS: Triple quadrupole. The triple quadrupole consists of two sets of parallel metallic rods inter-spersed by a collision cell. The first quadrupole scans the ions coming from the ionization source and allows only ions of a particular m/z ratio to pass through. These ions enter the collision cell where they are fragmented by collision against an inert gas like argon. The smaller fragments then enter the third quadrupole, which scans all the ions in the radio frequency mode to generate a spectrum based on the varying behavior of ions in an oscillating electrical field.

7.2.3 Q-TOF

Q-TOF combines the properties of both triple quadrupole and time of flight (Yang et al., 2021). The addi-tion of TOF to quadrupole increased the sensitivity and resolution of the instrument. The daughter ions so formed from the third quadrupole now move at a different speed as per their mass and hence resolve much better as compared to single TOF or TOF-TOF or TQ. Figure 7.4 shows a schematic for a Q-TOF design.

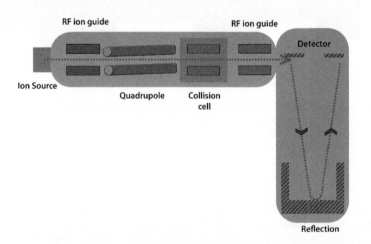

FIGURE 7.4   A schematic representation for a Q-TOF mass spectrometer.

7.2.4 LTQ-Ion Trap

In a LTQ-Ion trap, the daughter ions fed from the quadrupole are selectively eluted by the ion trap, maintained at a threshold potential difference by the ring electrodes. Thus, there is an increase in resolution, as the detector now faces fewer ions at a particular time. Orbitrap is a high-resolution mass analyzer with high mass accuracy, which is second stage of a hybrid mass spectrometer, whose first stage is typically a linear ion trap. The ions are injected into the orbitrap tangentially and form oscillating rings around the central electrode, becoming trapped in an electrostatic field.

7.2.5 LTQ-FTICR

The best resolution is provided by linear triple quadrupole (LTQ) coupled with ICR. The daughter ions released from the third quadrupole are focused toward the detector by the strong magnets placed between the triple quadrupole and the detector. As a result, the daughter ions keep resonating around the magnet before being sequentially eluted. However, ICR being extremely complicated, this configuration finds limited usage for clinical proteomics.

7.3 Comparison between Various Hybrid Configurations

For usage in day-to-day clinical proteomics, an MS configuration is necessary that provides high-throughput data as well as increased sensitivity and resolution. The reasons for such criteria are already discussed above. All the configurations discussed have different advantages and limitations. To address different biological questions or to analyze different samples, different configurations may be adopted. For example, analyzing complex cerebrospinal fluids, where potential biomarkers may be present at attomole range, Q-TOF or LTQ-orbitrap may be beneficial, whereas for multiple reaction monitoring for biomarkers or drug metabolism, TQ may be advantageous. Thus, the usage of a particular configuration is subject to the sample being processed for what objective. Nonetheless, all hybrid MS configurations provide strong analytical platform for highly reproducible and accurate data acquisition for proteomic applications.

REFERENCES

Glish, G. L., & Burinsky, D. J. (2008). Hybrid mass spectrometers for tandem mass spectrometry. *Journal of the American Society for Mass Spectrometry, 19*(2), 161–172. https://doi.org/10.1016/j.jasms.2007.11.013

Hopfgartner, G., Varesio, E., Tschäppät, V., Grivet, C., Bourgogne, E., & Leuthold, L. A. (2004). Triple quadrupole linear ion trap mass spectrometer for the analysis of small molecules and macromolecules. *Journal of Mass Spectrometry, 39*(8), 845–855. https://doi.org/10.1002/jms.659

McLafferty, F. W. (1981). Tandem mass spectrometry. *Science, 214*(4518), 280–287. https://doi.org/10.1126/science.7280693

Sleno, L., & Volmer, D. A. (2004). Ion activation methods for tandem mass spectrometry. *Journal of Mass Spectrometry, 39*(10), 1091–1112. https://doi.org/10.1002/jms.703

Suckau, D., Resemann, A., Schuerenberg, M., Hufnagel, P., Franzen, J., & Holle, A. (2003). A novel MALDI LIFT-TOF/TOF mass spectrometer for proteomics. *Analytical and Bioanalytical Chemistry, 376*(7), 952–965. https://doi.org/10.1007/s00216-003-2057-0

Yang, M., Li, J., Zhao, C., Xiao, H., Fang, X., & Zheng, J. (2021). LC-Q-TOF-MS/MS detection of food flavonoids: Principle, methodology, and applications. *Critical Reviews in Food Science and Nutrition, 0*(0), 1–21. https://doi.org/10.1080/10408398.2021.1993128

Exercises 7.1

1. The highest resolution is provided by which of the following instruments?
   a. Q-TOF
   b. TQ-FTICR
   c. TOF-TOF
   d. TQ

2. To increase the path length of ions in a TOF tube, _____ is used.
   a. Reflectron
   b. Refractron
   c. MALDI
   d. ESI

3. The first quadrupole in TQ is maintained at _____ mode.
   a. Scanning
   b. Radio frequency
   c. Operational
   d. None of the above

4. FT-ICR separates daughter ions using a constant _____.
   a. Electric field
   b. Magnetic field
   c. High voltage
   d. All of the above

5. The reflector in a TOF analyzer is used for?
   a. Compensation of minor differences in kinetic energy
   b. Increase the voltage
   c. Reduce the energy
   d. None of the above

6. Which of these is one of the most high-resolution mass analyzers that can resolve complex biomolecules and operates on the principle of frequency to dissolve ions?
   a. Triple quadrupole
   b. Fourier transformer
   c. Fourier transformer ion cyclotron resonance
   d. Electron-positron cyclotron collider

7. Mass of a tryptic peptide is 500 Da. What would be the m/z value of the ion [P+2H] +2? (P refers to the peptide)
   a. 500
   b. 502
   c. 250
   d. 251

8. Why do we integrate a separation technique (liquid chromatography) with a mass spectrometer?
   a. Helps in denaturing the protein
   b. Reduces sample complexity as all the peptides do not elute together
   c. Minimizes variations as all the peptides elute together
   d. Separates proteins on the basis of biological interactions

9. In a triple quadrupole mass spectrometer, which quadrupole acts as a collision cell?
   a. Q1
   b. Q2
   c. Q3
   d. Q4

10. Which one of the following processes is common for all mass spectrometry-based proteomic workflows?
    a. Liquid Chromatography
    b. Isoelectric Focusing
    c. 2-DE
    d. Proteolysis

11. Compare the various properties of the mass analyzers and state with reason which one of all the combinations would you use to detect the changes in the proteome profile of Stage II and Stage IV of a particular cancer. Note that, you don't need to mention about any particular protein, but the abundance of the proteins in Stage II and Stage IV are different.

Answers

1. b
2. a
3. a
4. b
5. a
6. c
7. d
8. b
9. b
10. d
11. Resolvability, sensitivity, and scan rate issues.

8

Tandem Mass Spectrometry for Protein Identification

Preamble

The use of mass spectrometry in the field of protein chemistry has greatly revolutionized the field and contributed as one of the major factors for the emergence of proteomics. Mass spectrometry was initially used to identify the molecules based on their mass. However, due to advancements in mass analyzers and more sophisticated hybrid configurations, identifying and quantifying proteins has become one of the major applications of this technique. The use of mass spectrometry in determining the sequences of proteins and also *de novo* sequencing (where the genome sequences are not known) has gradually popularized. Due to the modification of existing mass spectrometers by adding another mass analyzer and detector in tandem, it led to the tandem MS/MS analysis. In this chapter, we introduce the principle and importance of tandem MS/MS for protein identification.

8.1 Principle of Tandem Mass Spectrometry

Tandem mass spectrometry is a modified and advanced version of conventional mass spectrometry. A mass spectrometer has three components; ionization source (generating ions), mass analyzer (filtering ions) and mass detector (generating the signal) (McLafferty, 1981). Before we discuss new concepts, it would be useful to refresh few concepts related to typical proteomics experiments, in-gel digestion and LC-MS/MS. Most proteomics experiments involve the separation of protein mixture by means of electrophoresis followed by elution of the protein band of interest (in case of gel-based proteomics). This protein is then digested into small peptide fragments by means of proteolytic enzymes, where the most commonly used protease is trypsin. These small peptide fragments can then be further analyzed by MS. The peptide fragments obtained after digestion can be analyzed by MALDI-TOF or tandem MS/MS (Figure 8.1). In MALDI-TOF, peptide ions are accelerated at different velocities depending on their mass to charge ratios. The spectrum generated provides a set of peaks whose masses represent the peptides present in the mixture.

Mass Spectrometry analysis-Tandem MS/MS

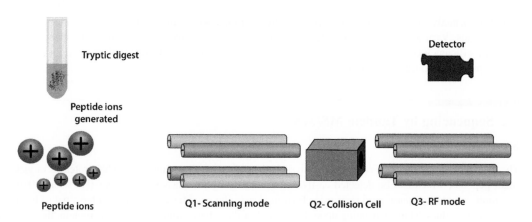

FIGURE 8.1 Illustration of a typical proteomics experiment using tandem MS.

DOI: 10.1201/9781003098645-11

Tandem MS/MS is capable of providing more in-depth sequence information. Each peptide in the digest is further fragmented in the second ionization step and analyzed, thereby generating a spectrum for each peptide. These spectra can then be analyzed by available software to obtain more information about the protein.

Tandem mass spectrometry involves more than one mass analyzers and detectors, kept in tandem with the primary mass analyzer. The linear triple quadrupole (TQ) is an excellent example of tandem mass spectrometry (Hopfgartner et al., 2004). From the previous chapter, you may recall that a TQ has three components, Q1 operating at the scanning mode, i.e., allowing specific ions to move into Q2. Q2 is the chamber where collision-induced dissociation (CID) takes place, and all the ions generating in Q2 move into Q3. Q3 or the third quadrupole operates in radio frequency (RF) mode, i.e., allowing all the ions, entering, to move toward the detector. You may also recall that the efficiency of TQ in terms of mass resolution and sensitivity and mass coverage was not very good. Therefore, hybrid mass configurations were developed having the basic TQ, associated with other high-end mass analyzers like TOF or ion trap or FT – ICR. This increases the sensitivity and the resolution power of the approach, and thus tandem MS/MS becomes useful for proteomics experiments. Different mass analyzers can be coupled with ionization source such as MALDI-TOF-based analysis (Figure 8.2) (Gogichaeva et al., 2007).

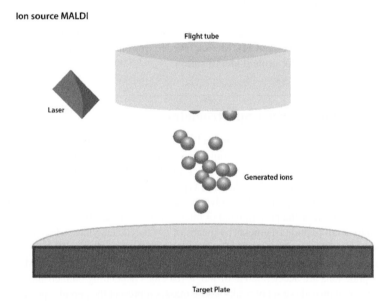

FIGURE 8.2   A schematic for MALDI-TOF-based mass spectrometry.

The mass analyzer resolves the ions produced by the ionization source on the basis of their mass-to-charge ratios. Various characteristics such as resolving power, accuracy, mass range and speed determine the efficiency of these analyzers. Commonly used mass analyzers include time of flight (TOF), Quadrupole (Q), ion trap and orbitrap.

8.2 Sequencing by Tandem MS/MS

8.2.1 b- and y-Ions

The first step for any mass spectrometric analysis is the digestion of proteins by suitable proteases to yield small peptides, which can be detected easily and accurately by the mass spectrometer (Harrison, 2009). The most common protease used is trypsin. Trypsin cleaves the peptides at the carboxyl-terminal of lysine or arginine residues. Thus, depending upon the amount of these basic amino acids in the protein, the number of peptides obtained post-tryptic digestion varies. The obtained peptides are ionized and accelerated towards the mass analyzer. The peptides on their trajectory collide with other peptides and break further. The fragmentation mainly takes place at the CO-NH or amide bond, thereby generating daughter ions, which are positively charged either at the carboxyl side or the amine side. If the positive charge is retained on the carboxyl side, it is known as the y-ion, whereas if it is retained on the amine side, it is called the b-ion (Figure 8.3).

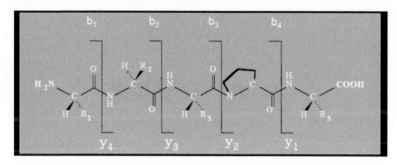

FIGURE 8.3 Peptide ion fragmentation and generation of b- and y-ions.

Further, the y-ions are more intense than the b-ions. Besides the b- and y-ions, several other types of ions are also generated (Figure 8.4), but those are usually not of significant importance when it comes to sequencing.

a. When the primary carbon and the amide carbon bond break, it gives rise to "x" ions (positive charge retained on amide carbon) and "a" ion (positive charge retained on primary carbon).

b. When the bond between amide nitrogen and primary carbon breaks, "z" ions (positive charge retained on primary carbon) and "c" ions (positive charge retained on amide nitrogen) are formed.

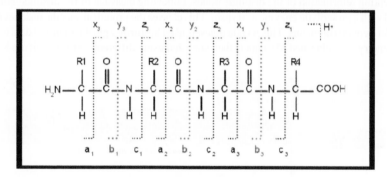

FIGURE 8.4 Various fragmentation patterns in tandem MS/MS.

8.2.2 Factors Affecting the Generation of b- and y-Ions

The generation of b-ions and y-ions is highly crucial for the generation of protein sequences. The generation of b-ions and y-ions takes place under low collision energy, because of the difference in the electronegativity of carbon and nitrogen. The generation of different types of ions is quite natural, but strategies are developed to enrich the b- and y-ions.

Certain amino acids in the peptide backbone affect fragmentation patterns. For example, unusual fragmentation by loss of water may occur in serine, threonine, hydrogen sulfide loss in cysteine, ammonia loss from glutamic acid and aspartic acid, etc. Proline, an imino acid, affects the fragmentation pattern at both the levels of proteolysis by trypsin and at the level of generation of b- and y-ions. When proline is present at the carboxyl-terminal after lysine or arginine, trypsin fails to cleave the peptides. This may be due to the cyclic nature of proline that inhibits the enzymatic action of trypsin. This problem can be mitigated using a different protease, like chymotrypsin, which cleaves at carboxyl-terminal of aromatic amino acids. Cleavage near aspartic acid or glutamic acid residues generates intense ions, because of the excessive charge retained on the daughter peptide, which affects the ability to read correct sequences.

Electrophoretic separation of a protein mixture results in distinct protein bands. These proteins can be used for analytical purposes by carrying out in-gel digestion (Figure 8.5). The protein spot/band in the gel are fragmented into small pieces with each piece being dissolved in a suitable buffer.

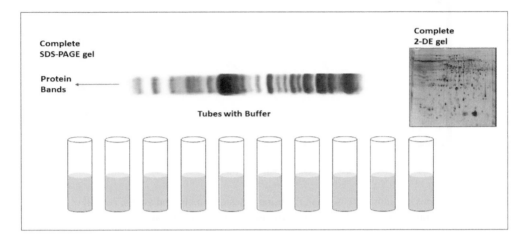

FIGURE 8.5 Schematic for in-gel digestion of protein.

The protein solution is treated with a reducing agent like dithiothreitol, which cleaves the disulfide bonds in the protein. This is followed by treatment with iodoacetamide, which alkylates the sulfhydryl groups thereby preventing reformation of the disulfide bonds. Following cleavage of the disulfide bonds, the protein is treated with a proteolytic enzyme as trypsin. This cleaves the protein at specific residues thereby generating smaller peptide fragments. This tryptic digest is used for further purification and data is acquired by mass spectrometry. The obtained MS data is then analyzed using different software (Figure 8.6).

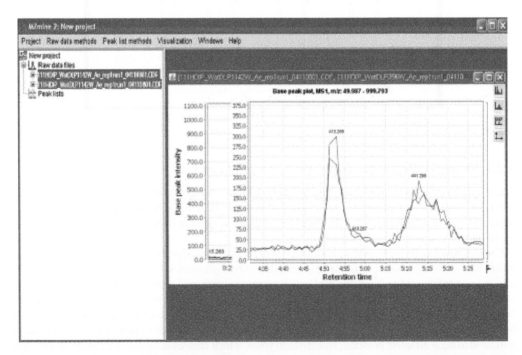

FIGURE 8.6 Methodology for LC-MS/MS data analysis. The methodology for the LC-MS/MS data analysis using representative software is illustrated in animation.

8.2.3 Deriving the Protein Sequence

A typical fragmentation results in b-ions and y-ions, which differ in one amino acid, just like automated DNA sequencing using dideoxy nucleotides, where the oligonucleotides differ from each other by one nucleotide. Identifying the amino acid from the difference in mass sequentially provides the sequence of the protein. The software (e.g., MASCOT) enables data analysis based on the mass of the various y-ions and b-ions. The difference between two successive y-ions or b-ions would yield the sequence of the amino acid (Figure 8.7).

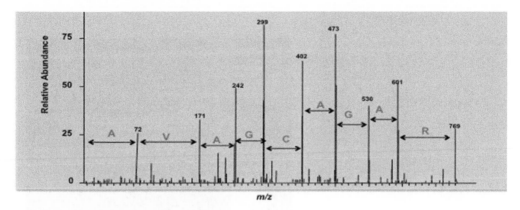

FIGURE 8.7 Deriving amino acid sequence from MS/MS spectrum.

The MS/MS data analysis shareware has some extra inputs such as quantitation, MS/MS tolerance, peptide charge, instrument, etc., in addition to the fields for peptide mass fingerprint (PMF). They require inputs from the user regarding the experimental parameters used such as enzyme cleavage, protein name, modifications, etc., and the desired search criteria like taxonomy, peptide tolerance, etc. Commonly used protein databases against which the MS information is processed to retrieve sequence data include NCBI, MSDB, and Swiss-Prot. The data file generated from MS is uploaded and the search is carried out. There are available software in which the results are obtained in a tabular format for the spectra. This information can be used for evaluating the sequence of peptide (Figure 8.8).

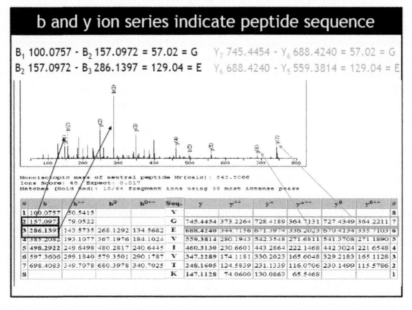

FIGURE 8.8 Generating sequences from b-ions and y-ions.

The tandem MS protein analysis is used to obtain protein identities from each of the sequenced peptides. The result-page begins with a list of probable protein identities and their respective sources. The score histogram provides details similar to the PMF analysis, with the probability distribution being displayed graphically. In Mascot, the green-shaded region is indicative of a match that has greater than 5% chance of being random while the red peak indicates that the chances of a random match are less than 5% (Figure 8.9).

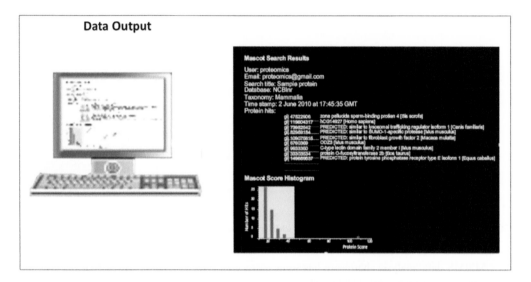

FIGURE 8.9   MS/MS data analysis.

The summary report lists all the protein matches obtained from the database search with their respective molecular weight, protein score, source organism, and details regarding each of its fragmented peptides. Further information about any of the protein sequences can be obtained by clicking on the corresponding protein link. Data related to each of the peptide fragmentation patterns can also be obtained by clicking on the peptide link indicated by the query number. The protein view obtained on selecting a particular protein link is very similar to the protein view observed in PMF. It provides details regarding the protein score, molecular weight, isoelectric point, the sequence coverage of the protein, etc. Protein scores above 67 are considered significant, and greater the percentage sequence coverage, more the number of matching peptides for that particular protein. All sequences are displayed with the matching sequences being indicated in red. Additionally, details regarding each peptide that matches are shown. Each peptide fragment's sequence, start and end amino acid locations, predicted and experimental molecular weights, number of missed tryptic cleavages, and accompanying ion scores are displayed. The final protein score is calculated using the highest ion scores.

Before reaching the detector in tandem MS/MS, each peptide goes through a second cycle of fragmentation in the second mass spectrometer. For each peptide fragment, this offers a substantially greater amount of information. Clicking on the peptide links in the summary report will take you there. By altering the x-axis plot values, the fragmentation pattern is shown graphically and can be zoomed in or out as needed.

Each peptide fragment is split at the amide bond under low collision energy, which can lead to the synthesis of the y-ion and the b-ion. By computing the mass difference between successive ions, the amino acid sequence can be determined using these ion masses. Each mass difference value is associated with a specific amino acid, which may be found in a common information table (Figure 8.10). As shown in the example above, the y-ion series and the b-ion series are in opposition to one another. The difference in the peaks' m/z ratios in Figure 8.10 shows the amount of mass lost during fragmentation. The loss of one amino acid is what caused this loss.

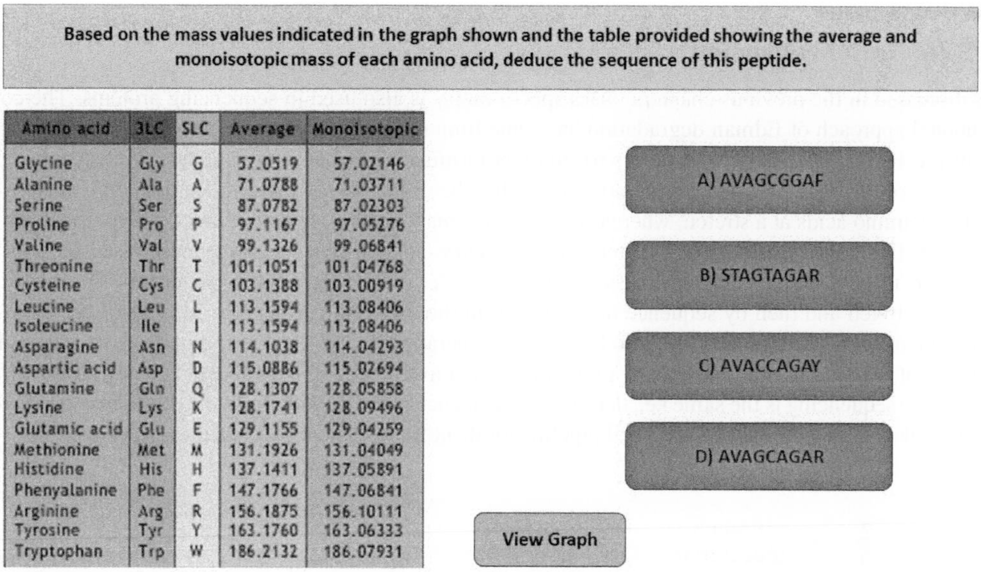

Based on the mass values indicated in the graph shown and the table provided showing the average and monoisotopic mass of each amino acid, deduce the sequence of this peptide.

Amino acid	3LC	SLC	Average	Monoisotopic
Glycine	Gly	G	57.0519	57.02146
Alanine	Ala	A	71.0788	71.03711
Serine	Ser	S	87.0782	87.02303
Proline	Pro	P	97.1167	97.05276
Valine	Val	V	99.1326	99.06841
Threonine	Thr	T	101.1051	101.04768
Cysteine	Cys	C	103.1388	103.00919
Leucine	Leu	L	113.1594	113.08406
Isoleucine	Ile	I	113.1594	113.08406
Asparagine	Asn	N	114.1038	114.04293
Aspartic acid	Asp	D	115.0886	115.02694
Glutamine	Gln	Q	128.1307	128.05858
Lysine	Lys	K	128.1741	128.09496
Glutamic acid	Glu	E	129.1155	129.04259
Methionine	Met	M	131.1926	131.04049
Histidine	His	H	137.1411	137.05891
Phenylalanine	Phe	F	147.1766	147.06841
Arginine	Arg	R	156.1875	156.10111
Tyrosine	Tyr	Y	163.1760	163.06333
Tryptophan	Trp	W	186.2132	186.07931

A) AVAGCGGAF

B) STAGTAGAR

C) AVACCAGAY

D) AVAGCAGAR

View Graph

FIGURE 8.10 Interactivity-derived peptide sequence.

Hint: You can use Table 8.1 for calculation.

TABLE 8.1

The Average and Mono-Isotopic Mass of Amino Acids

Amino Acid	3LC	SLC	Average	Monoisotopic Mass
Glycine	Gly	G	57.0519	57.02146
Alanine	Ala	A	71.0788	71.03711
Serine	Ser	S	87.0782	87.03203
Proline	Pro	P	97.1167	97.05276
Valine	Val	V	99.1326	99.06841
Threonine	Thr	T	101.1051	101.04768
Cysteine	Cys	C	103.1388	103.00918
Leucine	Leu	L	113.1594	113.08406
Isoleucine	Ile	I	113.1594	113.08406
Asparagine	Asn	N	114.1038	114.04293
Aspartic acid	Asp	D	115.0886	115.02694
Glutamine	Gln	Q	128.1307	128.05858
Lysine	Lys	K	128.1741	128.09496
Glutamic acid	Glu	E	129.1155	129.04259
Methionine	Met	M	131.1926	131.04048
Histidine	His	H	137.1411	137.05891
Phenylalanine	Phe	F	147.1766	147.06841
Arginine	Arg	R	156.1875	156.10111
Tyrosine	Tyr	Y	163.1760	163.06333
Tryptophan	Trp	W	186.2132	186.07931

8.3 *De Novo* Sequencing

As discussed in the previous chapters, mass spectrometry is also used in sequencing proteins. The conventional approach of Edman degradation has some limitations that are easily overcome by mass spectrometry. Firstly, Edman degradation is based on N-terminal sequencing of peptides, whereas there are no such constraints in mass spectrometry. Secondly, the efficiency of Edman degradation detection is up to 40 amino acids at a stretch, whereas the same for mass spectrometry in almost 100 amino acids at a stretch. In addition, the scan rate of mass spectrometry is much faster than the amino acid sequencer. However, in conventional tandem mass spectrometry, the mass spectrometer sequences several short peptides stretch and then by sequence homology from the database predicts the identity of the protein. But tandem mass spectrometry can also be used for generating *de novo* sequences of proteins, where the genome of the organism and hence the proteome are not available in the database. The overall principle of *de novo* sequencing is the same i.e., deriving the sequence of the peptide from the difference in mass of the peptides. Figure 8.11 shows a typical pipeline for de novo sequencing and data analysis.

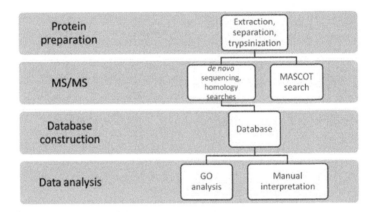

FIGURE 8.11   A typical workflow for *de novo* sequencing and analysis.

De novo sequencing finds its application in identifying novel splice variants and single nucleotide polymorphisms (SNPs), by the process of reverse genetics (Dančík et al., 1999). The sequences of proteins so obtained can be reverse read as the genome sequence by software after optimizing the degenerate genetic codes and codon biases in those organisms.

8.4 Challenges in Clinical Proteomics

Tandem MS/MS mainly finds its application in protein sequencing and *de novo* sequencing. It has become an integral part of mass spectrometry. In clinical proteomics, it is mainly used in biomarker discovery. But the area is still challenging, because validation is an important parameter for biomarker discovery, and the sample size also matters. The limitation in the field of clinical proteomics is partial because of instrumentations but mainly because of the pre-processing required for the samples and a large amount of sample that needs to be processed before it can be validated as a biomarker (Parker et al., 2010). In general, tandem mass spectrometry has revolutionized proteomics by virtue of high-resolution mass spectrometers.

REFERENCES

Dančík, V., Addona, T. A., Clauser, K. R., Vath, J. E., & Pevzner, P. A. (1999). De novo peptide sequencing via tandem mass spectrometry. *Journal of Computational Biology*, 6(3–4), 327–342. https://doi.org/10.1089/106652799318300

Parker, C. E., Pearson, T. W., Leigh Anderson, N., & Borchers, C. H. (2010). Mass-spectrometry-based clinical proteomics – A review and prospective. *Analyst, 135*(8), 1830–1838. https://doi.org/10.1039/C0AN00105H

Gogichaeva, N. V., Williams, T., & Alterman, M. A. (2007). MALDI TOF/TOF tandem mass spectrometry as a new tool for amino acid analysis. *Journal of the American Society for Mass Spectrometry, 18*(2), 279–284. https://doi.org/10.1016/j.jasms.2006.09.013

Harrison, A. G. (2009). To b or not to b: The ongoing saga of peptide b ions. *Mass Spectrometry Reviews, 28*(4), 640–654. https://doi.org/10.1002/mas.20228

Hopfgartner, G., Varesio, E., Tschäppät, V., Grivet, C., Bourgogne, E., & Leuthold, L. A. (2004). Triple quadrupole linear ion trap mass spectrometer for the analysis of small molecules and macromolecules. *Journal of Mass Spectrometry, 39*(8), 845–855. https://doi.org/10.1002/jms.659

McLafferty, F. W. (1981). Tandem mass spectrometry. *Science, 214*(4518), 280–287. https://doi.org/10.1126/science.7280693

Exercises 8.1

1. The number of peptide fragments generated from the sequence GAKTYFRPGVLIPKGL on trypsin digestion are?

    a. 3

    b. 2

    c. 4

    d. 1

2. y-ions are more intense than b-ions because?

    a. Charge is retained on carboxyl side

    b. Charge is retained on amino side

    c. Charge is retained on primary carbon

    d. Charge is retained on carbonyl oxygen

3. b-ions and y-ions are generated by?

    a. High-energy collision

    b. Low-energy collision

    c. Particle bombardment

    d. *De novo*

4. One of the crucial steps of sample preparation while subjecting it to mass spectrometry is in-gel digestion. This is the step in which protein spots/bands are excised from the gel and subjected to enzymatic digestion. Match the following reagents to the role that they serve during in-gel digestion and protein quantitation.

Reagent	Role
A) Acetonitrile	i) Alkylation
B) Ammonium bicarbonate	ii) Quantification
C) DTT	iii) Dehydration
D) IAA	iv) Digestion
E) Trypsin	v) Rehydration
F) iTRAQ	vi) Reduction

    a. A-iii, B-v, C-vi, D-i, E-ii, F-iv

    b. A-ii, B-v, C-vi, D-i, E-iv, F-iii

    c. A-iii, B-vi, C-v, D-i, E-iv, F-ii

    d. A-iii, B-v, C-vi, D-i, E-iv, F-ii

5. *De novo* sequencing can be used to detect SNPs based on?

    a.   Forward genetics

    b.   Reverse genetics

    c.   Top-down approach

    d.   Bottom-up approach

6. The proteolytic digestion with enzymes like trypsin is important to?

    a.   generate peptides with molecular weight in the range of mass spectrometry.

    b.   determine concentration of proteins.

    c.   remove DNA from sample.

    d.   remove stain from Coomassie stained gel bands.

7. Which of the following is *NOT* a part of mass spectrometry?

    a.   ESI

    b.   TOF

    c.   MALDI

    d.   SCX

8. A peptide mixture containing three different peptides was subjected to ESI-MS. Based on the information given below, choose the correct order of migration (first to last).

Peptide	Molecular Weight	Charge on the Peptide
P	1000	1
Q	2000	2
R	3000	3

    a.   P, Q, R

    b.   R, Q, P

    c.   Q, R, P

    d.   All three peptides will reach at the same time

9. Which of the following amino acids acts as a trypsin digestion breaker?

    a.   Histidine

    b.   Glycine

    c.   Proline

    d.   Lysine

10. The ionization source in mass spectrometry leads to ionization of the peptides. Which of the following parameters influence the deflection of ions in a mass analyzer?

    a.   Only the mass of the ion

    b.   Only the charge on the ion

    c.   Both the charge and mass of the ion

    d.   Neither the charge or mass of the ion

11. What are b-ions and y-ions? Derive the sequence of this peptide from both ends.

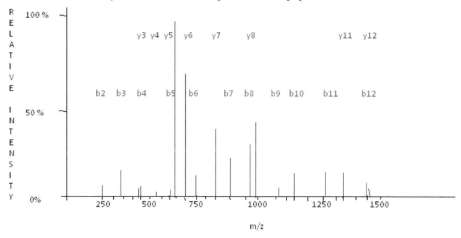

Answers

1. a
2. a
3. b
4. d
5. b
6. a
7. d
8. d
9. c
10. c
11. From the left-hand side b2 – b3 equivalents to a particular amino acid and from the right-hand side, y12 – y11 equivalents to the same amino acid.

  B2 – B3 = 250 – 340 = 90 = Proline
  Y12 – Y11 = 1450 – 1360 = 90 = Proline
  B3 – B4 = 340 – 450 = 110 = Leucine
  B4 – B5 = 450 – 610 = 160 = Tyrosine
  B5 – B6 = 610 – 750 = 140 = Histidine
  B6 – B7 = 750 – 900 = 150 = Phenylalanine
  Y8 – Y7 = 990 – 850 = 140 = Histidine
  B7 – B8 = 900 – 960 = 60 = Glycine
  Y7 – Y6 = 850 – 700 = 150 = Phenylalanine
  B8 – B9 = 960 – 1060 = 100 = Threonine
  Y6 – Y5 = 700 – 645 = 55 = Glycine
  B9 – B10 = 1060 – 1150 = 90 = Proline
  Y5 – Y4 = 645 – 540 = 105 = Threonine/Cysteine
  B10 – B11 = 1150 – 1260 = 110 = Leucine
  Y4 – Y3 = 540 – 445 = 95 = Proline
  B11 – B12 = 1260 – 1450 = 190 = Tryptophan

9

In Vitro *Quantitative Proteomics Using iTRAQ*

Preamble

One of the major advantages of mass spectrometry over conventional gel-based proteomics, apart from deciphering the identity of the protein, is the ability to quantify the proteins in the sample under consideration. The conventional mass spectrometry provides an idea about the relative abundance of the ions in question but using different isotopic or isobaric tags, it is now possible to determine the relative abundance. Although there are many quantitative proteomic approaches using mass spectrometry but isobaric tags for relative and absolute quantitation (iTRAQ), TMT and stable isotope labeling with amino acids in cell culture (SILAC) have become most popular. However, these techniques have their pros and cons and that's why these techniques are selected based on the type of applications. Amongst all these techniques, iTRAQ is currently one of the most popular techniques because of its versatility and multiplexing ability. This chapter mainly discusses the principle, working protocol and some advantages and disadvantages of iTRAQ-based quantitative proteomics.

9.1 Quantitative Proteomic Approaches

Proteins are the most dynamic entities that govern cellular activities. Hence it becomes necessary to study the dynamic change in protein concentration, subject to the change in various environmental, physiological or medical processes. Conventional techniques like enzyme-linked immunosorbent assay (ELISA) and radioimmunoassay (RIA) can be used to quantitatively study the dynamic changes in protein concentration but these techniques are limited to study only one protein at a time and also require the availability of pure antibodies against these proteins. Gel-based proteomic analysis has some disadvantages of reproducibility, membrane protein analysis, abundant protein representation, etc. Advanced gel-based proteomic analysis such as difference gel electrophoresis (DIGE) gives a qualitative and semi-quantitative analysis of change in proteome level, subject to various external stimuli.

In clinical proteomic studies, it becomes necessary to know the absolute difference in the levels of proteins subject to two different conditions, ideally, a normal/healthy against treated/pathological states. Also, the number of samples to be analyzed becomes extremely large to negate any false-positive results, especially when biomarker discovery is in question. For this high throughput requirement, mass spectrometry based quantitative proteomics has emerged as an essential tool in clinical proteomics (Hoofnagle, 2010; Ong et al., 2003). Mass spectrometric approaches include isotope-coded affinity tag (iCAT), isobaric tags for relative and absolute quantitation (iTRAQ), and stable isotope labeling with amino acids in cell culture (SILAC), which can quantitatively determine the difference at proteome level, subject to two or more conditions. The control and treated samples are labeled with different reporter tags. The difference in light and heavier reporter ions provides quantitative differences, and protein identity is also established simultaneously. The major advantage of iTRAQ technique is the availability of multiplexing, i.e., studying the difference in the samples of more than two patients (four or eight). This multiplexing helps in high-throughput analysis, reduces manual artifacts, and accelerates the proteomic analysis.

9.2 Principle of iTRAQ

iTRAQ is an *in vitro* labeling technique where the peptides are labeled with specific tags, which is used for quantitation during MS analysis (Wiese et al., 2007). Since MS is capable of measuring the mass (or m/z) of the peptide, the tags are so designed that for each sample the tag remains unique and at the same time for all the samples pooled together, the overall mass of various tags remains the same. The iTRAQ reagent contains a reporter group, which is lost during fragmentation at MS/MS step. This reporter group is different for different iTRAQ tags and they differ from each other by mass of 1 Da. Thus, when the peptide is identified by the first mass analyzer, it is moved into the second mass analyzer for further fragmentation, the reporter ion is released and is measured by the detector (Figure 9.1). Based on the number of samples, reporter ions are generated and hence their absolute quantitation is a measure of quantitation of peptide and therefore quantitation of protein in various samples. The advantage of iTRAQ is the ability to perform multiplexing, i.e., analyzing more than two samples at the same time. Currently, iTRAQ reagents are available with four or eight reporter groups and hence, at a time four (4-plex) or eight (8-plex) samples can be analyzed and quantified. Proteins from different samples are digested with trypsin, and peptides so obtained are differentially labeled with the iTRAQ reagents. The iTRAQ reagents interact with the primary amine or the N-terminal region of the protein and thus there is no bias for any particular amino acid, which is a limitation in some quantitative proteomic techniques.

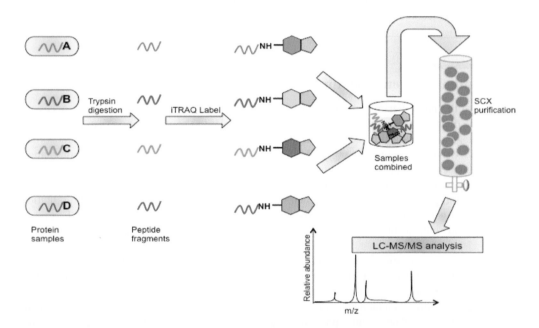

FIGURE 9.1   A generalized pipeline for isobaric tag for relative and absolute quantitation (iTRAQ) based proteomics.

9.3 iTRAQ Reagent

As explained earlier, iTRAQ is based on the labeling of peptides with isobaric tags, i.e., tags having the same molecular weights, but producing different ions during fragmentation, due to their inherent property. iTRAQ-based labeling depends on the availability of a primary amine and hence it is also called N-terminal labeling. The iTRAQ reagent consists of three regions (Figure 9.2).

a. A reactive group, which specifically binds to the primary amine group of the peptides. Chemically it is an ester of N-hydroxy-succinimide.

b. A balance group, which is chemically a carbonyl group.

c. A reporter group, which is chemically dimethyl piperazine.

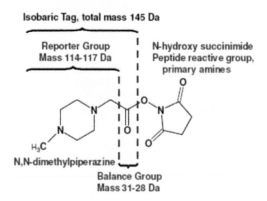

FIGURE 9.2   Structure of an iTRAQ reagent.

The reporter group differs in their weight but the overall weight of the tag remains constant due to the balance group. The mass is kept constant at 145 Da. In a 4-plex experiment (Figure 9.3), the reporter group has a mass ranging from 114 to 117 Da, and accordingly balancer group having mass ranging from 31 to 28, i.e., a reporter group having mass 114 Da has a balancer group of mass 31 Da, while a reporter group of mass 117 Da has a balancer group of mass 28 Da. The reactive group interacts with the primary amine of the peptide and forms a strong amide linkage, which retains even during fragmentation. The reporter group produces excellent signature ion and b- and y-ion series. It also maintains the ionization state of the peptide and hence better ion capture.

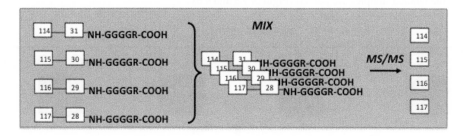

FIGURE 9.3   A typical 4-plexing iTRAQ experiment.

The identification and quantitation of complex protein mixtures have been facilitated by MS-based quantitative proteomic techniques. The iTRAQ reagent consists of amine-specific, stable isotope reagents that can label peptides from four to eight different biological samples. The protein samples to be analyzed are first digested with trypsin into smaller peptide fragments. The trypsin cleaves the proteins at the C-terminal of lysine and arginine residues unless they are followed by a proline residue. The limitations of iCAT regarding multiplexing were addressed by iTRAQ, where multiplexing could be done due to inherent properties of balancer and reporter group (Figure 9.4). This multiplexing was further increased to 8-plex, thus increasing the high throughput requirement in clinical proteomic studies.

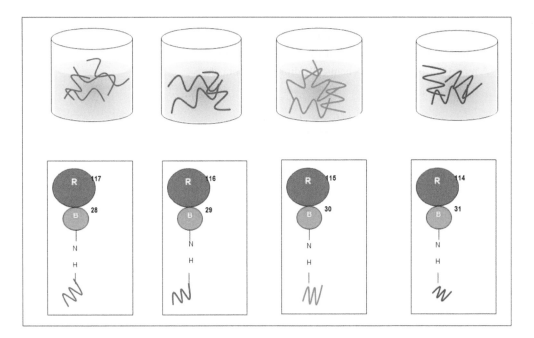

FIGURE 9.4   iTRAQ labeling showing masses of balancer and reporter groups.

9.4 Working Protocol

Mass spectrometry has its limitations in terms of sensitivity. Mass spectrometry is also sensitive to salt contamination and hence samples like cerebrospinal fluid or serum, containing a high concentration of salt need to be processed before MS analysis. The samples are first labelled with iTRAQ reagents and then all the labeled samples are pooled (Figures 9.5 and 9.6).

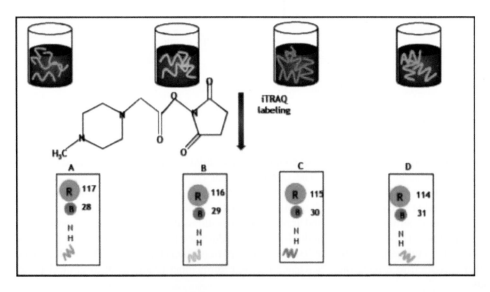

FIGURE 9.5   Peptide labeling by iTRAQ reagent.

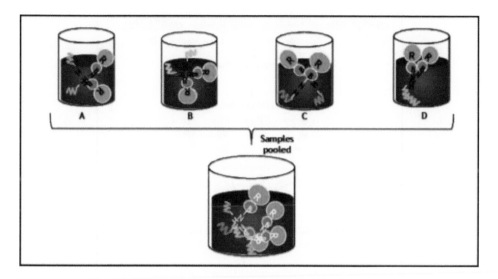

FIGURE 9.6 An overview of iTRAQ labeling where individual labeled samples are pooled.

The protein is dissolved in mass spectrometry compatible solvent, ammonium bicarbonate, pH 8.5 and then reduced to break all the disulfide bonds by using either dithiothreitol (DTT) or β-mercaptoethanol. To reduce, abnormal fragmentation from amino acids like cysteine, which gives out hydrogen sulfide ions, cysteine blocking reagents like methyl methane thiosulfonate (MMTS) is also added. The reduced protein is then trypsinized and salt contamination is removed by zip tipping. In a zip tipping process, the protein is passed through microtip columns containing C-18 material, which hydrophobically interacts with the protein and trap them, eluting out the salts and other contaminants. The protein is eluted back using acetonitrile and formic acid. The peptides so obtained are labeled with iTRAQ reagents and then pooled. The pooled samples are passed through strong cation exchange columns to remove excess iTRAQ reagent and obtain specifically labeled peptides prior to MS analysis (Figures 9.7 and 9.8) (Burkhart et al., 2011).

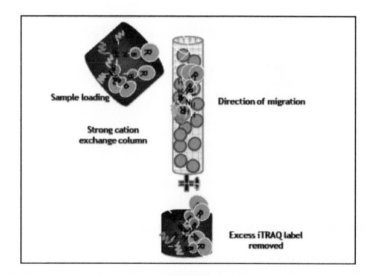

FIGURE 9.7 iTRAQ protocol: chromatography for clean-up and purification.

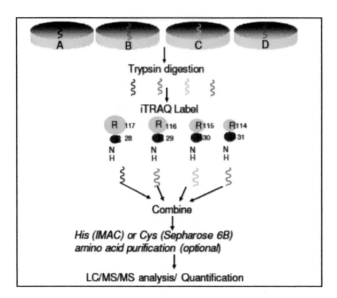

FIGURE 9.8   iTRAQ labeling for a 4-plex experiment.

This step facilitates sample clean-up prior to further finer separation and purification using reverse-phase chromatography. Further purification of the strong cation exchange (SCX)-purified peptides is carried out by reverse-phase liquid chromatography, wherein the sample is passed through a column containing a packed stationary phase matrix that selectively adsorbs only certain analyte molecules. The eluted fractions are further analyzed by MS. The purified labeled peptide fragments are then analyzed by MS/MS.

9.5 Data Analysis

iTRAQ enables simultaneous identification and quantification of peptides and hence proteins. Different samples are labeled using different tags as with balancer group of 191, 190, 189, 188 corresponding to reporter ion of 114, 115, 116, and 117, respectively (Figure 9.9).

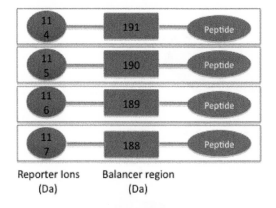

Reporter Ions   Balancer region
    (Da)   (Da)

FIGURE 9.9   Representing the masses of iTRAQ tags along with the balancer.

The raw data obtained after iTRAQ-MS is further analyzed to obtain the quantitative information. The chromatogram can also be analyzed to see the comparative peak intensity of different labeled samples. In Figure 9.10, the lower spectrum is the data obtained from the second MS analyzer, which identifies the protein and also quantifies it. For instance, the peak at 115.1643 is our peak of interest where absolute quantification data of our protein, obtained from four different samples, lies. Zooming into the spectrum at m/z 115.1643 provides finer details about the quantity of reporter ions generated from the peptide (shown in the top spectrum). The reporter ion corresponding to m/z 115.104 shows maximum intensity and that corresponding to m/z 114.110 shows the least.

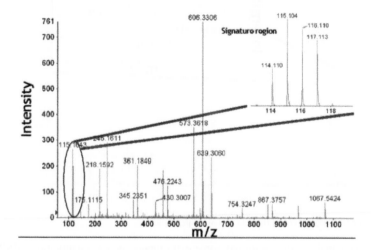

FIGURE 9.10 MS spectrum obtained after an iTRAQ 4-plex MS experiment.

This quantitative information becomes important for the validation of biomarkers. For example, in glioblastoma, certain biomarkers like haptoglobin or retinol-binding protein show varying concentrations at different stages of cancer. Quantifying the amount of the biomarker provides an idea about the grade of glioblastoma the patient is suffering from and hence can be diagnosed accordingly. Similarly, in 8-plexing, eight different samples are processed and labeled. The purified labeled peptide fragments are analyzed by MS/MS. The different masses of the reporter groups allow the peptide fragments to be identified. The reporter group is lost during fragmentation. Relative quantification samples can now be performed using iTRAQ. The reporter ion reports from 113.14 to 120.14 Da are considered and accordingly the data is analyzed (Figures 9.11 and 9.12).

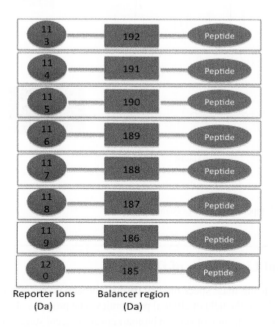

FIGURE 9.11 iTRAQ labeling for an 8-plex experiment.

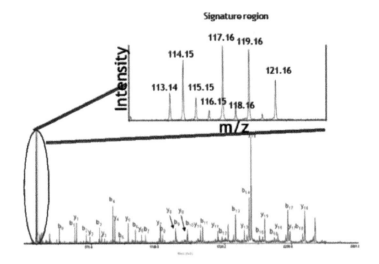

FIGURE 9.12   An 8-plex MS spectrum from an iTRAQ experiment.

The MS/MS data analysis shareware has some extra inputs such as quantitation, MS/MS tolerance, peptide charge, instrument, etc., in addition to the fields for PMF (Burkhart et al., 2011). They require inputs from the user regarding the experimental parameters used such as enzyme cleavage, protein name, modifications, etc., and the desired search criteria taxonomy, peptide tolerance, etc. Commonly used protein databases against which the MS information is processed to retrieve sequence data include NCBI (National Center for Biotechnology Information), and Swiss-Prot/UniProt. The data file generated from MS is uploaded and the search is carried out (Figure 9.13).

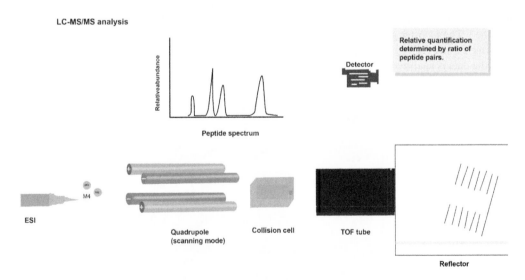

FIGURE 9.13   Data analysis pipeline for relative quantification.

9.6 Advantages and Limitations of iTRAQ

Quantitative proteomics using mass spectrometry-based approach is one of the most important advantages of iTRAQ, especially in the field of clinical proteomics. The progress of metabolism of a particular drug or the progressive concentration of certain biomarkers at different stages of a particular disease can easily be quantified using iTRAQ-based studies. Another great advantage of iTRAQ over iCAT is the independence of specific amino acids, like iCAT depends on cysteine labeling. Since the reactive group of iTRAQ reagent only interacts with a primary amine, there appears no dependence on any specific amino acid and hence no biasness occurs. This increases the amount of proteome covered by the technique, unlike iCAT. Multiplexing ability is another great advantage of iTRAQ (Ma et al., 2014). Since the detection is based on the reporter ion solely, the greater the number of different types of reporter ions, the greater is the number of samples that can be analyzed in one experiment. Currently, the number of reporter ions available is eight and hence, multiplexing at the level of eight samples is achievable. This becomes important in the discovery of biomarkers for clinical diagnosis, because a large number of samples need to be validated simultaneously.

However, iTRAQ reagents are extremely costly and also very sensitive to contamination from salts. Thus, extensive zip tipping and peptide enrichment need to be performed. Also, there is a requirement for sophisticated software for analyzing iTRAQ data. Another disadvantage of the entire protocol and not of iTRAQ, in general, is the variability arising due to the inefficient enzymatic digestion. Nonetheless, iTRAQ is a very powerful tool for the absolute quantification of multiple samples in an efficient manner.

REFERENCES

Burkhart, J. M., Vaudel, M., Zahedi, R. P., Martens, L., & Sickmann, A. (2011). iTRAQ protein quantification: A quality-controlled workflow. *PROTEOMICS, 11*(6), 1125–1134. https://doi.org/10.1002/pmic.201000711

Hoofnagle, A. N. (2010). Quantitative clinical proteomics by liquid chromatography–tandem mass spectrometry: Assessing the platform. *Clinical Chemistry, 56*(2), 161–164. https://doi.org/10.1373/clinchem.2009.134049

Ma, S., Sun, Y., Zhao, X., & Xu, P. (2014). Recent advance in high accuracy iTRAQ for quantitative proteomics. *Sheng wu gong cheng xue bao = Chinese Journal of Biotechnology, 30*(7), 1073–1082.

Ong, S.-E., Foster, L. J., & Mann, M. (2003). Mass spectrometric-based approaches in quantitative proteomics. *Methods, 29*(2), 124–130. https://doi.org/10.1016/S1046-2023(02)00303-1

Wiese, S., Reidegeld, K. A., Meyer, H. E., & Warscheid, B. (2007). Protein labeling by iTRAQ: A new tool for quantitative mass spectrometry in proteome research. *PROTEOMICS, 7*(3), 340–350. https://doi.org/10.1002/pmic.200600422

Exercises 9.1

1. Which of the following techniques is based on N-terminal labeling?
   a. iCAT
   b. iTRAQ
   c. SILAC
   d. DIGE

2. In iTRAQ, it is the _____ which is recorded for absolute quantification.
   a. Balancer group
   b. Parent ion
   c. Reporter group
   d. Reactive group

3. Zip tipping is done to?
   a. Enrich the peptides
   b. Remove salt contamination
   c. Both the above
   d. None of the above

4. MMTS is added to prevent anomalous fragmentation resulting from _____ residues?
   a. Lysine
   b. Threonine
   c. Cysteine
   d. Arginine

5. The advantage of iTRAQ over iCAT is?
   a. Multiplexing
   b. Independence of amino acid biasness
   c. Accuracy
   d. All of the above

6. What is the molecular weight of the balancer group in a 4-plex iTRAQ reagent?
   a. 187–191
   b. 28–31
   c. 190–193
   d. 27–30

7. What is the use of MS/MS spectrum during iTRAQ analysis?
   a. Biomarker detection
   b. Provides the amino acid sequence information
   c. Provides the sample information
   d. All of the above

8. Which of the following amino acids is involved in the iTRAQ-based labeling strategy?
   a. Only lysine
   b. Only arginine
   c. Both lysine and arginine
   d. Lysine and N-terminal amino acid

9. What is the mass range of the balancer group of an 8-plex iTRAQ reagent?
   a. 28–35
   b. 96–103
   c. 157–164
   d. 185–192

10. What is a reporter ion in an iTRAQ label?
    a. N-methyl piperazine
    b. Piperazine
    c. N,N-methyl piperazine
    d. N, N-dimethyl piperazine

11. Platelet Basic Protein is a 14 kDa protein found in the human serum. By using a quantitative MS approach, the change in PBP in response to benzene was analyzed as shown in the data. Explain what should have been the experimental strategy to obtain this data?

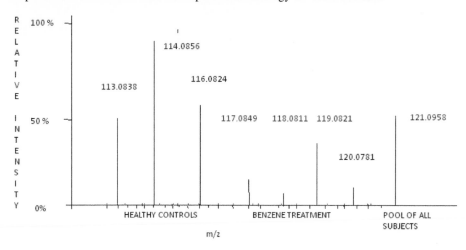

Answers

1. b
2. c
3. c
4. c
5. d
6. b
7. b
8. d
9. d
10. d
11. Platelet Basic Protein is a 14 kDa protein found in the human serum. The human serum contains abundant proteins like albumin, immune-globulin G and transferring, which constitute approximately 95% of all serum proteins. All these serum proteins are high molecular weight proteins having molecular weight above 50 kDa. Due to the abundance of these proteins the low molecular weight proteins, which can act as biomarkers for various disease diagnosis are masked. Hence, the strategy should be to deplete the human serum from these high molecular weight abundant proteins first. The experimental scheme can be as follows:

1. Collect serum proteins and desalt them (iTRAQ is sensitive to salt contamination).
2. Soluble form of the proteins is passed through affinity columns to remove IgG or serum albumin or they can be immune-precipitated.
3. Trypsinize them.
4. Take equivalent amounts of proteins from control and treated individuals and label them with the iTRAQ reagent. For example, labels with reporter ion of mass 113–115 can be mixed with healthy controls, 116–119 can be mixed with benzene treated samples and 120–121 can act as the internal standard.
5. Pass them through Strong Cation Exchangers to remove excess reagent and then elute them together. For better resolution, pass them through C14 resins and subject them to MS analysis.
6. Interpret the data by taking the ratio of the control peaks to the treated peaks of the reporter ions shown above.

10

In Vivo *Quantitative Proteomics Using SILAC*

Preamble

Mass spectrometry-based quantitative proteomics is an attractive platform for studying the dynamic changes in proteins subjected to various stimuli. The identification and quantitation of complex protein mixtures have been facilitated by MS-based methods for differential stable isotope labeling (Harsha et al., 2008). These tags, which can be recognized by MS, provide a basis for quantification. Stable isotope labeling by amino acids in cell culture (SILAC) incorporates specifically labeled amino acids into proteins for differential analysis.

Terminology

- **SILAC:** Stable isotope labeling of amino acids in cell culture – A technique for labeling the amino acids in cell culture medium, where proteins are labeled and then detected and quantified by MS.
- **ICAT:** Isotope coded affinity tag – A technique where cysteine residues of peptides are labeled and then quantified.
- **iTRAQ:** Isobaric tags for relative and absolute quantification – A technique where the N-terminal residues of the peptides are labeled followed by their detection and quantification using MS.
- **CDIT:** Culture-derived isotope tags – A technique where the cells are grown in different medium and then pooled together prior to MS analysis.

10.1 Limitations of iTRAQ and Emergence of *In Vivo* Labeling Methods

Mass spectrometry-based quantitative proteomics utilizes tags during the identification of proteins because of the specified difference in mass they generate. The tags can be introduced into the protein either by chemical, enzymatic, or metabolic routes. Chemical introduction of tags includes isotope-coded affinity tags (iCAT) and isobaric tags for relative and absolute quantitation (iTRAQ) where different isotopic or isobaric tags are mixed with trypsin-digested peptides and then the labeled peptides are analyzed (Sethuraman et al., 2004). The obvious disadvantage of this approach is that some peptides might escape the labeling of tag and hence remain undetected during identification and quantification. Also, these techniques generally depend upon specific amino acid residues, like cysteine for iCAT and primary amine for iTRAQ.

Enzymatic methods of incorporating tags include the usage of heavier oxygen in water during enzymatic digestion of the peptides. Trypsin hydrolyzes the peptide bond and in the process, adds the isotopically labeled water, thereby labeling the peptides. However, this approach does not give enough efficiency as iTRAQ. To address the above problems, metabolic means of incorporating tags were introduced by scientists, where the proteins are labeled *in vivo*. This can be achieved by labeling the media with contents required by the cells to synthesize proteins, like amino acids. Thus, the cells take up the labeled amino acids and hence the proteins formed are labeled. This approach has the benefit of not depending on any particular residue for the label and hence there is no biasness for any amino acid nor is the sensitivity of the instrument, because the protein becomes labeled uniformly.

10.2 *In Vivo* Labeling or Metabolic Stable Isotope Labeling

Stable isotope tagging methods use isotopic nuclei (e.g., ^{2}H, ^{13}C, ^{15}N, and ^{18}O) to determine the relative expression level of proteins in two samples. These methods use the biological incorporation of isotopically labeled nutrients or amino acids into proteins (Figure 10.1).

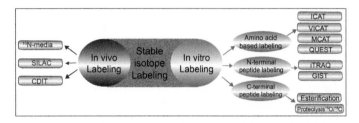

FIGURE 10.1   Stable-isotope tagging techniques based on *in vivo* and *in vitro* labeling for differential proteomics.

10.2.1 ^{15}N Labeling

In this method, yeast or bacterial cultures are grown in two separate media, one of which contains ^{15}N (Palmblad et al., 2007). Cells are pooled together, proteins are extracted and quantified on MS. One of the drawbacks associated with this method is a mass shift in the MS spectra. Since different proteins incorporate an unequal number of stable isotopes, labeled and unlabeled peptides exhibit a variable mass shift in the MS spectra. Moreover, this method is difficult and expensive for mammalian systems, which poorly incorporates stable isotope.

10.2.2 Culture-Derived Isotope Tags

For both relative and absolute quantitative proteome studies, in culture-derived isotope tags (CDIT) method, cells cultured in a stable isotope-enriched medium are mixed with tissue samples to serve as an internal standard. After protein extraction and separation, digested proteins are analyzed by MS to identify and quantify peptides. The ratio between the two isotopic distributions (from a tissue sample and isotope-labeled cells) can then be determined using MS.

10.2.3 Emergence of SILAC

The usage of stable isotopes of nitrogen for labeling proteins worked well with prokaryotic systems but mammalian systems would incorporate the tag poorly. Hence, the idea developed into providing isotope-labeled amino acids in the media for the growth of specific mammalian cells. This led to the emergence of stable isotope labeling by amino acids in cell culture (SILAC) for stable isotopic labeling of amino acids. As a result of SILAC, all proteins are in vivo incorporated with particular amino acids. It is used to detect differences using non-radioactive isotopic labeling.

10.3 Principle of SILAC

SILAC is a metabolic strategy of incorporating tags into proteins by the usage of isotope-labeled amino acids in the cell culture media (Chen et al., 2015). It depends on the cellular protein synthesis machinery for the uptake of amino acids from the media and incorporating them into the proteome. Ideally, in SILAC, a wide range of isotopes can be used. Mostly, ^{13}C is used as a stable isotope for labeling the amino acids. The usage of carbon as a stable isotopic tag was limited earlier due to the presence of carbon in other bio-molecules. In SILAC, due to the ability to specifically label amino acids, it becomes easier to use either nitrogen or carbon or oxygen as a stable isotopic tag.

As the cell grows in labeled lysine or arginine-containing media, slowly the cellular machinery starts incorporating the tags into the proteins, and with the natural turnover of the cell and proteins, ultimately after a few generations, the entire proteome is labeled. As a control, cells are also grown in light media for comparison (Figure 10.2). SILAC provides a basis for absolute quantification of proteins because the tags are naturally incorporated during protein synthesis and hence there is no chance of fabrication as is the case for iCAT or iTRAQ, where the labels are added externally after protein digestion. The difference in the relative abundance of peptides between heavy and light samples is a measure of the degree of differential expression of that particular protein on external stimuli.

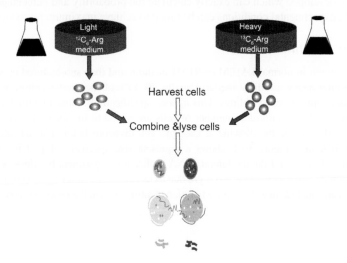

FIGURE 10.2 An overview of a SILAC-based experiment.

10.4 SILAC Experimental Workflow

A typical SILAC experiment involves several steps as given in Figure 10.3 including preparation of SILAC media, cellular adaptation, differential treatment, cell lysis and protein estimation, and data analysis (Ong & Mann, 2006).

1. Preparation of SILAC labeling medium
2. Adaptation of cells: from DMEM to SILAC labeling media
3. Differential treatment applied to the SILAC cells
4. Cell lysis and protein estimation
5. MS analyses and quantitation

FIGURE 10.3 Workflow of SILAC-based experiment.

10.4.1 Preparation of SILAC Media

The cell culture media used for SILAC strictly depends on the cells to be cultured. For most mammalian cells, Dulbecco's Modified Eagle Medium (DMEM) or Roswell Park Memorial Institute media (RPMI) are used. The media is prepared by usual formulations without adding the supplementary amino acids. At this stage, the pre-media is divided into two parts, in one part the natural amino acids are added, while in the other, isotopically labeled amino acids are added. Certain amino acids like lysine and arginine are isotopically labeled and added into the heavier medium, keeping all the other amino acids isotopically light. Both lysine and arginine have a backbone of six carbon atoms and hence provide a heavier mass of 6 Da to the labeled protein, which is the basis for quantification. Other components such as fetal bovine serum for growth factors, antibiotics to prevent contamination and other vitamins are added to make up the final media.

There is an inherent ability of the cellular machinery to recycle some of the amino acids. Sometimes, apart from synthesizing proline by the biosynthetic pathway, some amount of proline is also formed by the breakdown of arginine. This proline to arginine conversion is inevitable and at the same time diminishes the quality of the spectrum and hence hampers the data analysis. The problem can be mitigated by adding proline into the media. Still, some amount of arginine to proline conversion occurs. These days software has been developed which can exactly calculate the probability and percentage of this conversion. Accordingly, experimental conditions can be tuned to achieve maximum reproducibility.

10.4.2 Cellular Adaptation

The cells are first grown in normal DMEM or RPMI medium and then sub-cultured by splitting into two petri plates containing heavy and light media. Usually, 10–15% of the master culture is split into heavy and light media and kept for cell doublings. This is done specifically because mammalian cells are prone to contact inhibition and so a minimum number of cells are added to the media to allow the stable integration of tags into the protein by allowing at least five cell divisions before contact inhibition stops cell division. The given figure (Figure 10.4) shows a representative spectrum of protein isolated from the fifth generation of cell division. Like the famous DNA replication experiment by Messelson and Stahl, the time when spectrum becomes consistent and high, due to incorporation and recycling of the labeled amino acids, it is then we can conclude that stable incorporation of amino acid in the entire proteome has taken place.

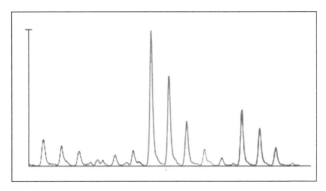

FIGURE 10.4   MS spectrum for cellular adaptation.

10.4.3 Differential Treatment

SILAC and other quantitative proteomic approaches were developed to analyze the dynamic variation in protein concentration due to some external stimuli. The external stimuli can be the effect of some drugs that need to be tested or some other physiological conditions that the protein might face. Differential treatment in SILAC is only provided when all the proteins have stably incorporated the tags into them. This cannot be manually checked but using MS it can be ensured that after four to five generations all the proteins are tagged.

10.4.4 Cell Lysis and Protein Estimation

Cell lysis is done by any normal procedure, specific to the cell line in question. Enzymatic lysis is preferred over any harsh method. The proteins are isolated, precipitated and then reduced using dithiothreitol (DTT) or beta-mercaptoethanol. Iodoacetamide (IAA) is added to alkylate the reduced disulfide bonds and to prevent re-oxidation of the bonds. Subsequently, proteins are then trypsinized, desalted by zip tipping and analyzed by mass spectrometry.

10.4.5 MS Data Analysis

The peaks so obtained after MS analysis are identified by searching databases. The relative quantification or fold change in mass between the normal and treated cells is 6 Da. Hence, two closely associated peaks differing in 6 Da are invariably from the same protein (Figure 10.5). Since the properties of the peptides are the same, they co-elute and hence are co-detected. The ratio of abundance of two peaks is a measure of the amount of fold change that has occurred due to differential treatment.

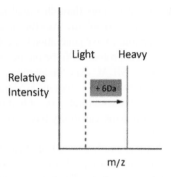

FIGURE 10.5 MS spectrum obtained from an SILAC experiment.

A typical workflow for a SILAC experiment is shown in Figure 10.6. Further, multiplexing in SILAC is also possible with the usage of the wide variety of substitutions in the arginine. For example, 5-plexing using arginine has been made possible in SILAC. The five different forms of arginine are:

a. Arginine with all the four Nitrogen labeled.
b. Arginine with all the six Carbon labeled.
c. Arginine with all the six Carbon and four Nitrogen labeled.
d. Arginine with all the six Carbon, four Nitrogen, and seven Hydrogen labeled.
e. Unlabeled arginine.

These forms of arginine vary from each other by a mass of 4, 6, 10 and 17 Da, respectively. The difference in the protein at subsequent levels is possible by taking the ratio of the various peaks.

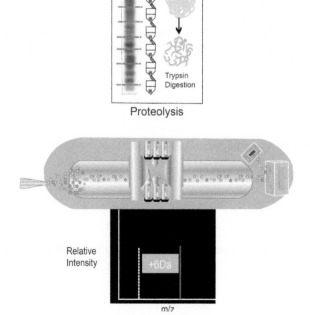

FIGURE 10.6 A typical workflow of SILAC.

10.5 Advantages and Limitations of SILAC

SILAC is far more effective at incorporating tags than chemical techniques like iCAT and iTRAQ. There is 100% incorporation of tags because SILAC involves the steady integration of tags into proteins through natural means. However, at least four to five generations of cell divisions are required to guarantee that the tags are added into the complete proteome. To be on the safe side, proteome analyses of each generation are also required. This makes a SILAC experiment more expensive and time-consuming.

Since the cells are cultured in the same medium, variations are less likely because only the medium's amino acids change. Additionally, there is reduced likelihood of handling errors because the samples are pooled prior to quantification. The multiplexing capability has also raised the level of high-throughput protein quantification (Yates et al., 2009).

The main drawback of SILAC is that it can only use cells that can be cultured. Serum or other bodily fluids cannot be evaluated by SILAC for this reason. This restricts the application of SILAC in clinical proteomics, particularly in the search for biomarkers. The requirement for full integration of amino acids in the proteome for analysis, unlike radioactivity, is another drawback of SILAC. Despite these drawbacks, SILAC is a very useful instrument and finds excellent uses in a variety of in vivo applications.

REFERENCES

Chen, X., Wei, S., Ji, Y., Guo, X., & Yang, F. (2015). Quantitative proteomics using SILAC: Principles, applications, and developments. *PROTEOMICS*, *15*(18), 3175–3192. https://doi.org/10.1002/pmic.201500108

Harsha, H. C., Molina, H., & Pandey, A. (2008). Quantitative proteomics using stable isotope labeling with amino acids in cell culture. *Nature Protocols*, *3*(3), 505–516. https://doi.org/10.1038/nprot.2008.2

Ong, S.-E., & Mann, M. (2006). A practical recipe for stable isotope labeling by amino acids in cell culture (SILAC). *Nature Protocols*, *1*(6), 2650–2660. https://doi.org/10.1038/nprot.2006.427

Palmblad, M., Bindschedler, L. V., & Cramer, R. (2007). Quantitative proteomics using uniform 15N-labeling, MASCOT, and the trans-proteomic pipeline. *PROTEOMICS*, *7*(19), 3462–3469. https://doi.org/10.1002/pmic.200700180

Sethuraman, M., McComb, M. E., Huang, H., Huang, S., Heibeck, T., Costello, C. E., & Cohen, R. A. (2004). Isotope-coded affinity tag (ICAT) approach to redox proteomics: Identification and quantitation of oxidant-sensitive cysteine thiols in complex protein mixtures. *Journal of Proteome Research*, *3*(6), 1228–1233. https://doi.org/10.1021/pr049887e

Yates, J., Ruse, C., & Nakorchevsky, A. (2009). Proteomics by mass spectrometry: Approaches, advances, and applications. *Annual Review of Biomedical Engineering*, *11*, 49–79. https://doi.org/10.1146/annurev-bioeng-061008-124934

Exercises 10.1

1. The amino acid used predominantly for labeling in SILAC experiments is?
   a. Arginine
   b. Proline
   c. Valine
   d. Leucine

2. The minimum number of generations required for stable integration of amino acids is?
   a. 5
   b. 8
   c. 3
   d. 6

3. IAA is added to
   a. Oxidise the disulfide bond
   b. Reduce the disulfide bonds
   c. Alkylate the disulfide bonds
   d. None of the above

4. In CDIT, the _____ acts as an internal standard
   a. Proteins from cells grown in light medium
   b. Synthetic peptides
   c. Proteins from cells grown in heavy medium
   d. None of the above

5. What is a disadvantage of using the SILAC method?
   a. Applicable for only cultured cells
   b. Labeled amino acids affect the protein synthesis
   c. Used for protein quantification
   d. 100% isotope incorporation

6. Which of the following is an *in vivo* protein labeling method used for protein quantitation using mass spectrometry?
   a. 15N media
   b. SILAC
   c. CDIT
   d. All of the above

7. Many protein labeling methods are available for mass spectrometry-based protein quantitation. Which of the following labeling methods that is also the first gel-free quantitative labeling method, depends on cysteine residue for labeling of proteins?
   a. iTRAQ
   b. TMT
   c. iCAT
   d. SILAC

8. Which of the following labeling methods involves N-terminal labeling?
   a. SILAC
   b. GIST
   c. iCAT
   d. All of the above

9. Which of the following is an *in vivo* labeling method used for mass spectrometry-based proteomics studies?
   a. iCAT
   b. SILAC
   c. QUEST
   d. GIST

10. Which of the following media is commonly used for SILAC labeling?
    a. RPMI
    b. MEM
    c. Fetal bovine serum
    d. All of the above

11. Yeasts are highly versatile organisms capable of tolerating wide range of environmental fluctuations. The signaling cascade in response to environmental fluctuations is a key to genetic manipulation of the organism.

    The yeast, *Saccharomyces cerevisiae* was subjected to hyper-osmotic stress and the variable change in the protein phosphorylation was studied to draw a comparison of the signaling cascade. Using SILAC-based approach, the role of two of the key players were studied, data of which is shown below. Note that Pan1 and Hog1 are kinases, while PTN9 and SOS3 are normal proteins. Please write your interpretation for the data shown below.

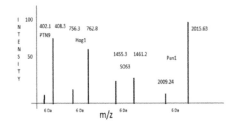

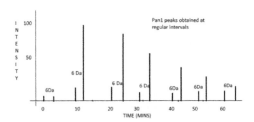

Western Blot Image of Pan1
at different time intervals

Western Blot Image of Pan1
at different time intervals

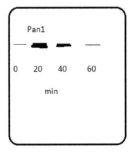

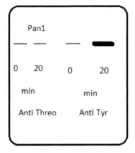

Answers

1. a
2. a
3. c
4. a
5. a
6. d
7. c
8. b
9. b
10. a

11. The following conclusions can be drawn from the following data.

    1. On hyper-osmotic stress, kinases Hog1 and Pan1 become phosphorylated and hence activated. There is also an over-expression of the protein PTN9 whereas SOS3 has no significant fold change.

    2. Time kinetic data of Pan1 suggests that the phosphorylation of Pan1 is transient, maximizing at 10 min and falling to the basal level at the end of an hour.

    3. Western Blot validation of Pan1.

    4. To check whether the phosphorylation of Pan1 is at the threonine residue, western blotting was done. It suggests that the phosphorylation does not occur at threonine residues but at the tyrosine residues.

On the basis of the above conclusions the following conclusions can be drawn:

• Hyper-osmotic stress activates Hog1 kinase. It is given that SOS3 and PTN9 are normal proteins. Given that SOS3 is not regulated whereas PTN9 is highly up-regulated, it can be hypothesized that Hog1 kinase must have a dimerization domain (like receptor tyrosine kinases), which can recruit PTN9 molecules. These PTN9 molecules are responsible for the hyper-activation of Pan1 kinase.

Module IV

Interactomics: Basics and Application

11

Introduction to Interactomics

Preamble

Proteins are the most important biomolecules of all biological entities. They are involved in every aspect of the cell structure and functioning as backbones for support, cell signaling, hormones, enzymes, mediators of the immune system, and as key players in almost every metabolic pathway of the cell. Despite their wide reach in all the cellular processes, the factor that is arguably more than the protein itself in the cell functioning is the interaction between two proteins. Without the correct interaction between the right proteins, no biological function can take place. Conversely, a single incorrect interaction or aberrant interaction could prove to be disastrous for the cell and possibly for the whole organism. This is one of the most fundamental characteristics in biology, be it for any metabolic pathway in the heart, brain, immune system, or even at any evolutionary level right from simple bacteria to plants to humans. Thus, the study of protein-protein interactions is one of the most pertinent topics for the understanding of biological functions. Another imperative reason why protein interactions are important to study is because they can cause major differences from normal to disease conditions. The excess or absence of a single metabolite might set off a chain of downstream processes, which eventually lead to an abnormal, diseased state. Thus, for achieving the crucial goal of understanding the disease, it is necessary to understand cell-signaling pathways involved in it and how the proteins interact under normal and abnormal conditions.

Terminology

- **Interactome:** Network of all the interactions in a cell. Interactomics comprises the study of interactions and their consequences between various proteins as well as other cellular components. The network of all such interactions, known as the "Interactome", aims to provide a better understanding of genome and proteome functions.
- **Yeast two-hybrid system (Y2H):** This is a novel molecular biology technique that is used for screening and discovery of protein-protein or protein-DNA interactions. In Y2H system, two proteins of our interest are tagged to a binding domain (BD) and an activation domain (AD) of the yeast transcription factor. Indirect interaction between the BD and AD also leads to transcription of the reported gene, thus proving interaction between the two proteins of interest.
- **Immunoprecipitation:** Immunoprecipitation or tandem affinity purification (TAP) is a technique that is used to purify protein complexes and study their protein-protein interactions. Depending on the protein that needs to be purified, different tags can be attached to the bait protein.
- **Protein microarrays:** These are miniaturized arrays normally made of glass, polyacrylamide gel pads, or microwells, onto which small quantities of many proteins are simultaneously immobilized.

11.1 Interactomics and Its Significance

Interactomics is the study of a network of interactions. Interactome consists of all the interactions among biological pathways and associated molecules. Interactomics utilizes a bioinformatics approach along with the experimental data. The necessity for the usage of bioinformatics can be attributed to the immense amount of information obtained for each network, the key players, their upstream/downstream interactors and the type of interaction, etc. However, in all the interactions, the most active players are the proteins as almost all major steps in all the pathways are mediated by proteins in the form of enzymes, hormones, receptors and metabolites etc. (Figure 11.1). Therefore, to understand the mechanism of cellular molecular processes, studying the protein-protein interaction becomes most essential.

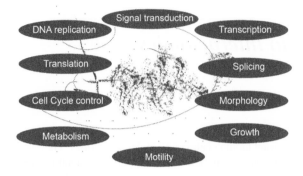

FIGURE 11.1   Protein-protein interactions: the potential cellular processes which may be affected due to improper protein-protein interactions.

There are different ways in which these interactions take place. These interactions can be strong or weak, or they may be transient or permanent and may finally result in the formation of homo-oligomers or hetero-oligomers. Any alteration in these interactions could lead to the formation of new, incorrect oligomers or render a key factor inactive, or a change in the kinetics of a certain reaction.

Each interactome is highly controlled and tightly regulated. Any imbalance can hamper normal conditions and result in a disease-like state. Thus, studying the protein interactions is important as they interact with a wide variety of biomolecules such as lipids, nucleic acids and small drug inhibitors, etc. Proteins also interact with one another to form macromolecular complexes that regulate signal transduction and gene regulation. The study of these networks could also help in understanding the function of uncharacterized proteins and also to find out new roles for characterized proteins. Moreover, new networks of protein interactions can be found out as the same protein may play a role in different pathways. Mechanisms to regulate protein activity can also be discovered, and this comprehensive knowledge will help provide an accurate picture of the real biological interactions. This understanding will further help us to manipulate the networks, thereby opening new avenues to study diseases.

Interaction studies of proteins with various biomolecules help in deciphering and understanding the functions of various proteins in the complex network of cellular pathways (Phizicky et al., 2003). Proteins interact with other biomolecules such as nucleic acids, lipids, hormones, etc., to perform a multitude of functions in living organisms such as signal transduction, growth and regulation, and metabolism, to mention a few. Protein interactions with other biomolecules can be of several different types (Figure 11.2).

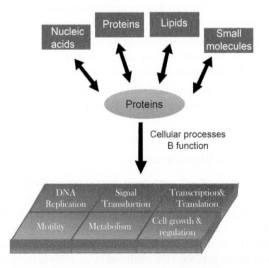

FIGURE 11.2   Different interactions studied in interactomics and their significance.

They may be weak or strong, obligate or non-obligate, transient or permanent. The physical basis for these interactions includes electrostatic, hydrophobic, steric interactions, hydrogen bonds, etc.

11.2 Methods to Study Protein-Protein Interactions

11.2.1 Traditional Approaches

11.2.1.1 Yeast Two-Hybrid System

The yeast two-hybrid system was developed by Fields and Song (Fields & Song, 1989; Fields & Sternglanz, 1994). The yeast two-hybrid system helps in the detection of protein-protein interaction by utilizing a transcription factor involving DNA-binding domain (DNA-BD) and an activator domain (AD). The main principle which enables this system to be used for studying protein-protein interaction is that the transcription factor has a modular nature and hence the transcription activation can occur even when the two domains are weakly bound, i.e., indirectly connected. In yeast two-hybrid system, the yeast transcription factor is divided into two separate parts, the DNA-BD and the AD. The two proteins of interest whose interaction is desired to study are bound to either domain. This is accomplished by generating genetically engineered plasmids which express a fusion protein of BD and the desired protein known as the bait. A second plasmid is engineered which expresses a fusion protein of the AD with the other protein to be studied termed as the prey.

The bait-BD and prey-AD are jointly expressed in a yeast cell. If the bait and prey interact, they will bind and bring the BD and the AD in proximity, which results in the activation of transcription of the reporter gene (Figure 11.3). If the bait and the prey do not interact, the BD and AD will not interact and the reporter gene transcription will not take place. Thus, one can find out whether two proteins interact with each other. This approach can also be used to screen a library of proteins.

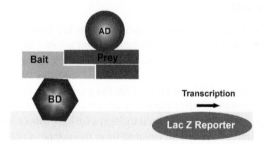

FIGURE 11.3 Yeast two-hybrid system: interaction between AD and BD leading to transcription of reporter genes.

The two-hybrid screening protocol uses this interaction as the basis for screening for protein interactions. When the bait protein bound with the binding domain interacts with the prey protein fused with the activation domain, there will be the expression of the reporter gene, which can easily be detected. LacZ is a commonly used reporter gene whose protein product is β-galactosidase which cleaves the substrate X-gal resulting in blue color. The binding of transcriptional activator proteins (composed of binding domain and activation domain) to the promoter region is essential for expression of the corresponding reporter gene located downstream of the promoter. The binding domain is fused with the bait protein, while the activation domain is fused with the prey protein. The binding of either one of the fusion proteins to the promoter is not sufficient to bring about transcription of the gene (Figure 11.4).

11.2.1.2 Affinity Chromatography

One of the most robust methods of detecting and confirming protein-protein interactions is the purification of the actual multi-protein complex by the use of affinity chromatography. For affinity purification, plasmids are generated which tag the protein of interest, resulting in a fusion product of the protein with the tag, either at the N-terminus or the C-terminus. Such plasmids are grown in bacterial cells, resulting

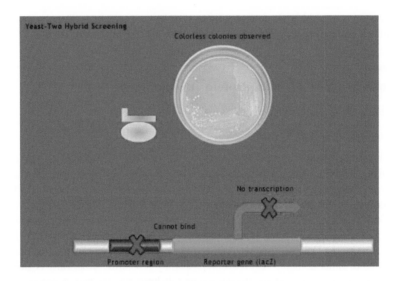

FIGURE 11.4   Representation of yeast two-hybrid screening.

in the production of the tagged protein. The protein is then purified using an antibody specific for that tag such as His (histidine), FLAG, GST (glutathione S-transferase), and Halo. After purification, the target protein and its interacting partners become isolated from the sample.

11.2.1.3 Immunoprecipitation

Using an antibody against the target protein, protein is extracted from the lysate in immunoprecipitation. In a co-immunoprecipitation, the known protein precipitates alongside the proteins that are associated to it, such as the protein's binding partners or the other components of a protein complex. As a result, it is feasible to identify novel, undiscovered interactors for the protein of interest. Tandem affinity purification (TAP) is an additional method for the purification of protein-protein interactions. Here, calmodulin-binding peptide (CBP) from the N-terminal, tobacco etch virus protease (TEV protease) cleavage site, and protein A are synthesized to form a fusion product with a TAP tag. The protein of interest and its various interactors can be isolated from the sample since the TAP tag binds to IgG robustly. Complexes are precipitated from the lysate and then subjected to polyacrylamide gel electrophoresis (PAGE), which divides the multi-protein complex into its component parts (Figure 11.5).

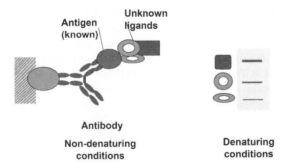

FIGURE 11.5   Affinity purification of proteins: in this illustration, a bait protein is tagged with affinity tags. Upon passage of this fusion protein through the affinity column, it binds to the affinity column. The elution of this fusion is done by adding another solution, which has a higher affinity for the column. Upon collection of the eluted protein sample, the protein separation is performed using SDS-PAGE and further validation and characterization of the protein are done using MS-based techniques.

The gel is excised for each protein band that has been separated. Following further processing, the removed gel pieces are evaluated using mass spectrometry, which aids in the identification of the complex's unidentified proteins.

The protein of interest is fused with a TAP tag, which contains a CBP, TEV cleavage site, and protein A. Depending upon the proteins to be studied, this tag can be modified. The tag is then bound to a column through affinity interactions between the protein A and IgG. The protein mixtures whose interactions with the bait protein are to be studied are then added. Some of the proteins form a complex with the bait protein through specific binding interactions. The remaining unbound proteins are then washed away. This is followed by cleavage at the TEV site by the TEV protease to release only the protein complex bound to CBP. These reactions constitute the first affinity step (Figure 11.6).

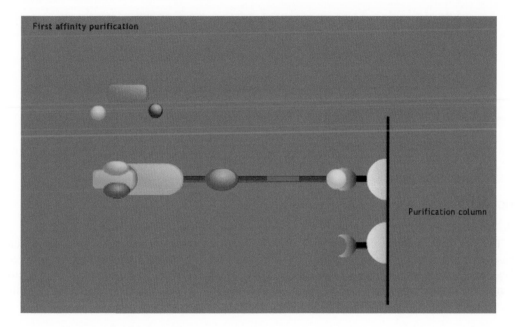

FIGURE 11.6 Schematic for steps in affinity purification.

In the second affinity purification step, the bait-prey complex is bound via the CBP domain to a calmodulin functionalized column in the presence of calcium ions. The column is washed to remove any other unwanted contaminants after which a chelating agent is added, which complexes the calcium ions. Once these are removed, the CBP-calmodulin interaction is weakened and leads to the release of the purified protein complex. Once the protein complex has been purified, the components of the complex are separated by electrophoresis under reducing conditions. The SDS gel is then analyzed and the protein components are evaluated thereby providing an understanding of the interactions with the bait protein of interest.

Once the unbound proteins are washed off the array surface, the protein interactions are detected by means of an array scanner. These protein microarrays are extremely useful in studying interactions with other proteins as well as small molecules, DNA, or RNA.

11.2.2 High-Throughput Techniques

11.2.2.1 Protein Microarrays

Although the traditional methods of studying protein-protein interactions are effective, they are very time-consuming thus, limit the number of proteins to be studied at a time. For studying a large number of proteins simultaneously, high-throughput methods are needed. One of the high-throughput methods developed for studying protein-protein interaction is protein microarrays (Chan et al., 2004). It is one of the most convenient methods for studying a large number of proteins in a single experiment (Zhu et al., 2001). Protein microarrays have a solid surface like a chip or a glass slide, on which thousands of proteins are immobilized. The array is incubated with specific probe molecules, which are tagged with a reporter molecule such as a fluorescent chromophore. The array is scanned using a laser or any appropriate method of detecting the reporter, and the interaction can be identified as the probes bind at the appropriate proteins' spots (Figure 11.7). In this way, thousands of targets can be screened in one go. This high-throughput platform can be used for biomarker discovery, cancer cell profiling (Sreekumar et al., 2001), antigen-antibody studies, studying protein-protein interactions, identification of new interactions, and functional characterization of interacting partners.

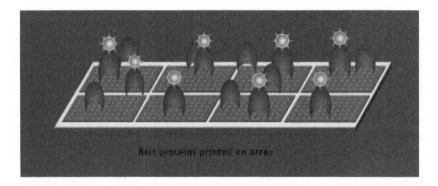

FIGURE 11.7   Bait proteins printed on the microarray.

The protein microarrays can be broadly classified into two kinds, abundance-based and function-based protein microarrays (Figure 11.8). The abundance-based protein microarrays consist of direct label-based, sandwich, and reverse-phase protein microarrays. While the function-based protein microarrays consist of peptide fusion, nucleic acid programmable protein array (NAPPA) and multiple spotting technique (MIST).

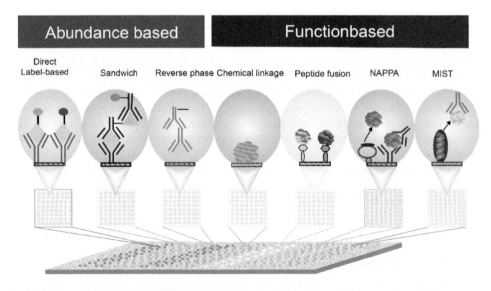

FIGURE 11.8   Various types of protein microarrays.

11.3 Applications

The protein-protein interaction studies can be used for various biological applications such as:

- Identification of novel proteins.
- Identification of function of unknown proteins.
- Identification of new binding partners for known proteins.
- Functional characterization of protein interactions.
- Protein/antibody screening.
- Biomarker discovery.

11.4 Challenges

Despite various distinctive advantages, the techniques involved in interactomics studies have several challenges as summarized in the Table 11.1.

TABLE 11.1

Advantages and Challenges of the Interactomics Techniques

Technique	Advantages	Challenges
Yeast two-hybrid	• Ability to screen large libraries • Simple protocol • No expensive equipment required • Can be used for DNA-protein, RNA-protein and protein-protein interactions • Easy to find out gene sequence • Highly sensitive for detection	• High false-positive and false-negative rates • Proteins must localize and interact in nucleus • Application in a non-yeast environment is questionable • Sensitive to toxic gene • Limited to pair-wise interaction • Difficult to detect interactions due to post-translational modifications
Affinity chromatography	• Proteins in native state • Interactions are natural • Large order complexes can be observed	• Sticky proteins appear regularly • Unclear whether interaction is direct or indirect • Expensive • Additional purification may lead to loss of weakly interacting proteins
Protein microarrays	• Large number of proteins can be assayed in a single experiment • Very small quantities of protein required • Easy screening of the arrays • Protein interaction with nucleic acids, lipids etc., can also be studied	• Proteins need to be purified • Protein functionality must be maintained even after binding to the array

11.5 Conclusions

In conclusion, interactomics is a powerful tool to understand the actual interaction networks in the cell thus helps in understanding various cellular pathways. It helps to identify aberrations in protein interactions which are linked with diseases.

REFERENCES

Chan, S. M., Ermann, J., Su, L., Fathman, C. G., & Utz, P. J. (2004). Protein microarrays for multiplex analysis of signal transduction pathways. *Nature Medicine*, *10*(12), 1390–1396. https://doi.org/10.1038/nm1139

Fields, S., & Song, O. (1989). A novel genetic system to detect protein–protein interactions. *Nature*, *340*(6230), 245–246. https://doi.org/10.1038/340245a0

Fields, S., & Sternglanz, R. (1994). The two-hybrid system: An assay for protein-protein interactions. *Trends in Genetics*, *10*(8), 286–292. https://doi.org/10.1016/0168-9525(90)90012-U

Phizicky, E., Bastiaens, P. I. H., Zhu, H., Snyder, M., & Fields, S. (2003). Protein analysis on a proteomic scale. *Nature*, *422*(6928), 208–215. https://doi.org/10.1038/nature01512

Sreekumar, A., Nyati, M. K., Varambally, S., Barrette, T. R., Ghosh, D., Lawrence, T. S., & Chinnaiyan, A. M. (2001). Profiling of cancer cells using protein microarrays: Discovery of novel radiation-regulated proteins. *Cancer Research*, *61*(20), 7585–7593.

Zhu, H., Bilgin, M., Bangham, R., Hall, D., Casamayor, A., Bertone, P., Lan, N., Jansen, R., Bidlingmaier, S., Houfek, T., Mitchell, T., Miller, P., Dean, R. A., Gerstein, M., & Snyder, M. (2001). Global analysis of protein activities using proteome chips. *Science*, *293*(5537), 2101–2105. https://doi.org/10.1126/science.1062191

Exercises 11.1

1. Which of the following is commonly used as an antibody specific tag for protein purification?
   a. His tag
   b. GST tag
   c. Halo tag
   d. All of the above

2. Which of the following is true about the yeast two-hybrid system?
   a. The false-positive rate is low and false-negative rate is high
   b. The false-positive rate is high and false-negative rate is low
   c. Both statement a and b are false
   d. Both statement a and b are true

3. Which of the following entities are bound to each other in the yeast two-hybrid system?
   a. Bait-AD
   b. Bait-CD
   c. Prey-AD
   d. Prey-BD

4. Which is the best method to study protein interactions with other macromolecules?
   a. Yeast two-hybrid system
   b. Protein microarrays
   c. Immunoprecipitation
   d. Affinity chromatography

5. The network of interacting elements in a cell is known as the _____.
   a. Proteome
   b. Metabolome
   c. Genome
   d. Interactome

6. A unique advantage of a yeast two-hybrid (YTH) system as compared to any of the other studied conventional techniques is to?
   a. Screen large clone libraries
   b. Application in non-yeast background is easily possible
   c. It can be used to study interactions beyond binary reactions in a single experiment
   d. Protein folding is guaranteed to be similar to that of mammalian systems

7. Which of the following is *not* an example of abundance-based protein microarrays?
   a. Direct label-based method
   b. Sandwich method
   c. Reverse-phase protein microarrays
   d. Peptide fusion

8. Which of the following is an advantage(s) of using protein microarrays?
   a. Large number of proteins can be assayed in a single experiment
   b. Very small quantities of protein required
   c. Protein interaction with nucleic acids, lipids etc. can also be studied
   d. All of the above

9. Which would be the best method for screening a large number of antibodies?
   a. Yeast two-hybrid system
   b. Immunoprecipitation
   c. Immunofluorescence
   d. Protein microarrays

10. In the yeast two-hybrid system, the AD domain binds to the bait domain, which leads to successful transcription system?
    a. True
    b. False

11. Given below are the components involved of a method, which is used for studying protein-protein interactions. Which method is being described here? Give a schematic representation of such interaction.

DNA-binding domain

Activation domain

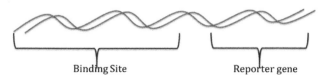

Binding Site   Reporter gene

DNA strand

Answers

1. d
2. c
3. c
4. b
5. d
6. a
7. d
8. d
9. d
10. b
11. The experiment which is used in this scenario to study protein-protein interaction is the yeast two-hybrid system. In this method, the DNA-binding domain (BD) binds to the appropriate DNA sequence. This BD acts as domain and the prey-activation domain (AD) attaches to this bait. The interaction between BD and AD activates the transcription of the reporter gene and is depicted in the figure below.

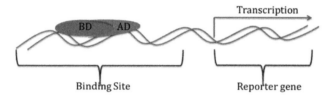

12

Antigen and Antibody Microarrays

Preamble

All the biological processes are mediated by the macromolecules in the cell and the most dynamic and versatile of these are the proteins. Proteins are extensively involved in every cell process including receptors, hormones, enzymes and metabolites, etc. Thus, for a complete understanding of the cellular processes, it is important to understand how the proteins interact with each other and other macromolecules. Traditional approaches such as yeast two-hybrid systems, immunoprecipitation, immunofluorescence, etc., have been in use for studying protein interactions. However, a very limited number of proteins can be studied at a time, making these approaches very time-consuming. For obtaining a large amount of data at the same time, high-throughput techniques have been developed recently. One of the high-throughput techniques used for studying protein-protein interaction is microarrays. Thousands of proteins can be screened simultaneously, thus combining many experiments in a single one. There are many ways in which the antibodies can be used to bind the antigen. Also, there are various methods for the detection of antibodies. Thus, depending on the type of experiment, antigen or antibody microarrays can be used for the detection of the antigen-antibody reaction.

Terminology

- **Protein microarrays:** A concept that evolved from DNA microarrays and provides a valuable platform for high-throughput analysis of thousands of proteins simultaneously.
- **Protein purification:** Since the target protein is expressed along with other proteins native to the host system, it is essential to purify the desired protein before printing it onto the array surface. For this reason, the gene of interest is often fused with a convenient tag sequence such as His6 that will facilitate the purification process.
- **Array functionalization:** The microarray surface must be suitably derivatized with a chemical reagent that can react with the groups present on the protein surface in order to firmly immobilize them on the microarray. Functionalization is often done with silane derivatives as these react easily with the groups present on the protein. Aldehyde groups react with amine groups present on the protein to form Schiff's base linkages, which hold the protein firmly in place.
- **SERS:** Surface-enhanced Raman Scattering – used for enhancing the electrical properties around the surface of the particle which can be detected using Raman scattering.
- **SWNTs:** Single-walled carbon nanotubes – SERS-based methods for detection of antigen-antibody reactions

12.1 Microarrays

Microarrays are platforms on which a large number of proteins can be studied at the same time (Chan et al., 2004). They make an important tool for studying protein function and abundance (LaBaer & Ramachandran, 2005; MacBeath, 2002). The setup includes solid support like a glass slide on which proteins are immobilized. The microarray slides are usually treated with amines, aldehyde or epoxy in order to facilitate the immobilization of the protein molecules. The sample is then incubated with a probe, which is linked to a label. The probe matches the appropriate target in the sample and a label can be used to detect the spot. This approach enables simultaneous screening of hundreds of proteins on a

DOI: 10.1201/9781003098645-16

single chip. Thus, it is a high-throughput method for screening a large number of targets at the same time. There are many microarrays depending on the nature of the proteins being captured on the array and different detection platforms as label-based and label-free (Figures 12.1 and 12.2).

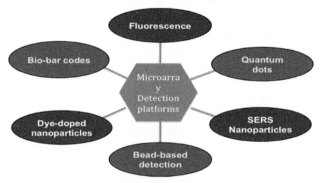

FIGURE 12.1   Label-based platform for microarrays.

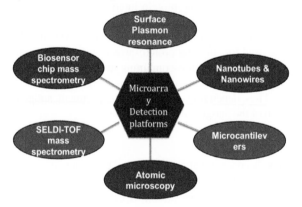

FIGURE 12.2   Label-free platform for microarray-based analysis.

Functional analysis of proteins is a time-consuming process that requires many steps. Analysis of a single protein at a time would be a tedious and laborious process. Analysis of several protein samples will undoubtedly take a long time if run one at a time. Protein microarrays successfully overcome (MacBeath, 2002) this hurdle by allowing analysis of several samples simultaneously (Figure 12.3).

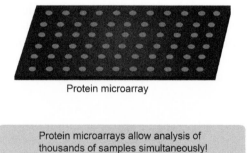

Protein microarray

Protein microarrays allow analysis of
thousands of samples simultaneously!

FIGURE 12.3   Need for protein microarrays.

The gene coding for the protein of interest is expressed in a suitable heterologous host system such as *Escherichia coli* by means of expression vectors like plasmids (Figure 12.4). The host cell machinery is used for transcription and translation resulting in a mixture of proteins consisting of the target protein and other host proteins. Since the protein of interest is expressed along with other proteins native to the host, it is essential to purify the target protein before it can be used for microarray applications.

Chromatographic procedures can do this to obtain the pure target protein. Protein purity is tested on SDS-PAGE gels. Tags like His6 are often fused with the protein of interest to facilitate the purification due to its specific affinity toward Nickel.

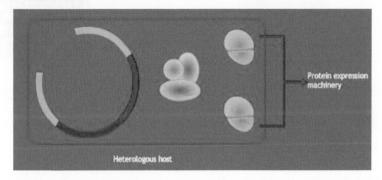

FIGURE 12.4 Protein expression and purification.

Commonly used array surfaces include glass, gold, nitrocellulose, and hydrogels (Figure 12.5). While glass slides are easy to handle, available at low cost, and can be used with existing scanning equipment. They show relatively low surface absorption and need to be derivatized with more reactive groups. They however continue to be used more extensively than the other array surfaces which are more expensive. Comparative features for commonly used array surfaces are demonstrated in the table. The array surface is functionalized with a suitable chemical reagent that will react with groups on the protein surface. Aldehyde and silane derivatization are commonly used as they interact well with amino groups present on the protein surface resulting in firm protein capture.

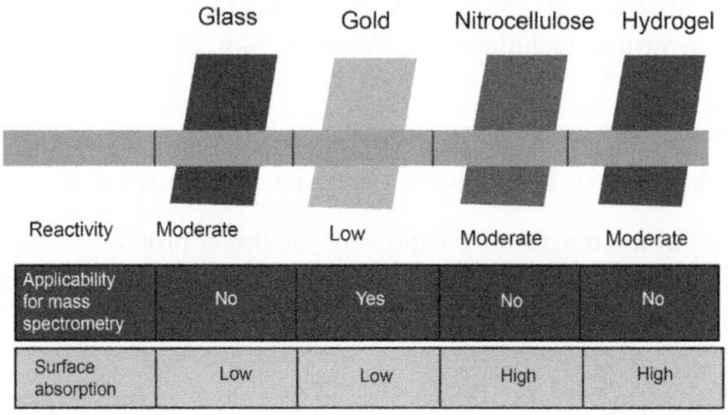

	Glass	Gold	Nitrocellulose	Hydrogel
Reactivity	Moderate	Low	Moderate	Moderate
Applicability for mass spectrometry	No	Yes	No	No
Surface absorption	Low	Low	High	High

FIGURE 12.5 Commonly Used Array surfaces.

The protein solution is printed onto the array surface in minimal volumes using a robotic printing device with small pins (Figure 12.6). The slides are kept for a suitable duration following the printing step to allow capture of the protein onto the array surface. The unreacted sites are then quenched by a blocking solution such as bovine serum albumin (BSA), which prevents any non-specific protein binding in subsequent steps.

There are two types of protein arrays that are commonly used; forward phase and reverse phase. In forward phase arrays, the analytes of interest, such as an antibody or aptamer are bound to the array surface and then probed by the test lysate that may contain the antigen of interest. In reverse-phase arrays, however, the test cellular lysate is immobilized on the array surface and then probed using detection antibodies specific to the target of interest. In the direct labeling detection technique, all the target proteins are labeled with a fluorescent or radioactive tag that facilitates easy detection upon binding to the immobilized capture antibody on the array surface. In the sandwich assay, however, a fluorescent-tagged secondary antibody that recognizes a different epitope on the target antigen binds to it and is detected through fluorescence. The protein microarray is then scanned in a microarray scanner that allows detection of the

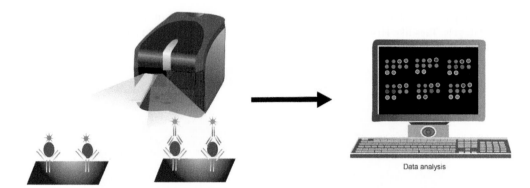

FIGURE 12.6   Protein array printing and functionalization.

fluorescently labeled proteins or antibodies. The output from this scanner is then received by software after which the data can be analyzed. Protein microarrays have found wide applications for discovery and functional proteomic studies. They allow rapid analysis of thousands of proteins simultaneously.

12.2 Antigen-Antibody Binding

Antigen-antibody reaction is one of the most specific interactions in biochemistry. Just like a lock and key, the antibodies have a very high affinity for the corresponding antigens and the commonly occurring reaction is very strong. The antibodies are very diverse and they can be easily produced for almost any protein. The antigen and its antibody are held together by non-covalent bonds such as hydrogen bond and Van der Waals force thus making the reaction reversible. Hence, the antigen-antibody reaction is used to detect the proteins in various techniques like Western Blot, ELISA, and microarrays, etc.

12.3 Antigen-Antibody Reactions Used in Microarrays

Antibodies are used in microarrays for capturing the target proteins or for reading the signal from the experiment (Brennan et al., 2010; Sanchez-Carbayo et al., 2006). The antigen-antibody interaction can be used in an array in multiple ways.

12.3.1 Direct Labeling

For a direct labeling procedure, the microarray surface is immobilized with the antibodies. Upon incubation with the sample, the target antigens bind to the appropriate antibody on the microarray (Figure 12.7).

Merits

- Only a single antibody is required.
- High reproducibility.
- Highly sensitive for abundant proteins.
- Multiple sample assay.

Demerits

- Less sensitivity for low abundant proteins.
- Chemically modified sample.
- Cross-reactivity.

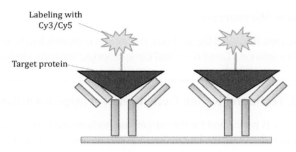

FIGURE 12.7   Direct labeling detection for microarrays.

12.3.2 Sandwich Assay

In sandwich assay, the antibodies are spotted on the microarray. The array is then incubated with the sample and the target antigen bind to the appropriate antibody on the array. A second antibody is then introduced which binds to the antigen, sandwiching the antigen between two different antibodies (Figures 12.8 and 12.9).

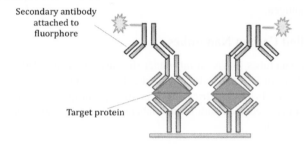

FIGURE 12.8   Indirect labeling detection for microarrays.

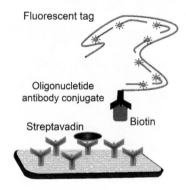

FIGURE 12.9   Rolling circle amplification for microarrays.

Merits

- Higher specificity.
- Very sensitive.

Demerits

- Cross-reactivity.
- Multiplexed analysis is not possible.
- High cost.

12.3.3 Reverse-Phase Microarrays

In reverse-phase microarrays, instead of the antibody, the antigen is immobilized onto the microarray plate. It is then incubated with a solution of antibodies and the appropriate antibody binds to the specific antigen.

12.4 Label-Based Technologies for Detection of Antigen-Antibody Microarrays

Once the microarray assay is performed for the antigen-antibody interaction, one needs to use an appropriate detection method for detecting the interaction. Various techniques are available for detecting antigen-antibody binding, and most involve a different secondary antibody. A few of these antibody-based detection techniques are listed below.

12.4.1 Fluorescence-Based Techniques

In fluorescence-based techniques, direct or indirect labeling is performed. In indirect labeling, the target antigen is labeled with a fluorescent dye, which then binds to the capture antibody on the microarray surface. In indirect labeling or sandwich technique, the unlabeled target antigen binds to the capture antibody on the microarray surface. A second antibody that is fluorescently tagged is introduced, which then binds the target antigen.

12.4.2 Single-Walled Carbon Nanotubes

Single-walled carbon nanotubes (SWNTs) fall under the surface-enhanced Raman scattering (SERS) type of detection method. In SERS detection methods, the Raman dye is coated to the particle, along with the antibody. The SWNTs have electrical and spectroscopic properties, with a characteristic Raman signature. These SWNTs are coupled with Raman-labeled antibodies, which make the detection of the target possible (Figure 12.10).

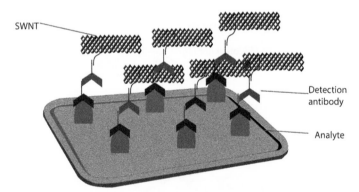

FIGURE 12.10   Representation of a single-walled carbon nanotube (SWNT) for microarray detection.

Merits

- High sensitivity.
- Multiplexed detection.
- Minimum background signal.
- Resistance to photo bleaching.

Demerits

- Metal impurities interfere with the activity.
- Insoluble in biological buffers.
- Difficult to determine the degree of purity.

12.4.3 Gold Nanoparticles

The gold nanoparticles work on the principle of surface plasmon resonance. The gold nanoparticles are attached to the antibody. Upon binding of the antibody to the antigen, a change in the emission spectra of the gold nanoparticles is observed which allows the detection of the interaction (Figure 12.11).

Merits

- Improved optical property.
- Superior quantum efficiency.
- Compatible with wide range of wavelengths.
- Resistance to photo bleaching.

Demerits

- High cost.
- Cytotoxicity.
- Non-uniform size and shape of nanoparticles.

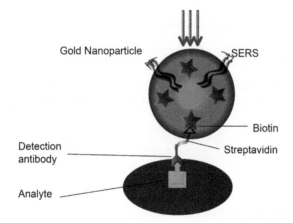

FIGURE 12.11   Gold nanoparticles for microarray detection.

12.4.4 Quantum Dots

Quantum dots are crystals with a semiconductor fluorescent core coated with another semiconductor shell. When light of higher frequency falls on the quantum dot, it leads to excitation. As the quantum dot returns to the lower energy level, the energy is released, which can be read as emission spectra (Figure 12.12).

Merits

- Brighter fluorescence.
- Excellent photostability.
- Multicolor fluorescent excitation.
- Greater quantum yield.

Demerits

- Toxicity.
- Unknown mechanism.

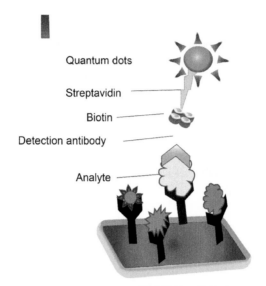

Antibody microarray

FIGURE 12.12   Quantum dots for microarray detection.

12.4.5 Dye-Doped Silica Nanoparticles

The dye-doped silica NPs consist of a silica matrix in which fluorescent molecules are packed. It can be tagged to various targets including antibodies and the interaction can be detected using fluorescence (Figure 12.13).

Merits

- Biocompatible.
- High sensitivity.
- Minimal aggregation and dye leakage.
- Photostability.
- High capacity.

Demerits

- Complex synthesis process.

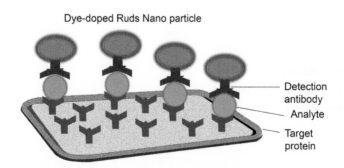

FIGURE 12.13   Dye-doped silica nanoparticles for microarray detection.

12.4.6 Magnetic Nanoparticle Probe Bio-Barcodes

In magnetic nanoparticle probe bio-barcodes, the nanoparticle probes are coated with DNA unique to the protein of interest. The magnetic microparticle probes (MMPs) consisting of antibodies for the target analytes, are captured by corresponding antibodies. This complex is then separated in presence of a magnetic field, and oligonucleotides are dehybridized and sequenced for the identification of protein of interest. The identification of the liberated oligonucleotide barcodes can be done on the microarray surface using either scannometric detection or conventional fluorophores (Figure 12.14).

Merits

- High sensitivity.
- Short detection times.
- Easy adaptability to multiple targets.

Demerits

- Can be used only with known antibodies.

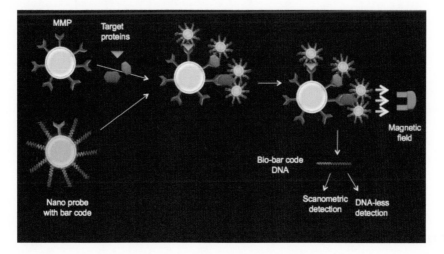

FIGURE 12.14 Magnetic nanoparticle probe bio-barcodes for microarray detection.

12.5 Applications

The antigen and antibody microarrays are useful in various applications as given below.

- Biomarker discovery.
- Autoantibody detection.
- Protein-protein interaction studies.
- Cytokine detection.
- Discovery of small-molecule inhibitors for signal transduction.
- Protein screening.
- Antibody screening.

There are available studies which employed different microarray approaches to understand protein interactions. Haab et al. (Haab et al., 2001) printed six arrays of 114 different antibodies onto poly-L-lysine-coated glass slides using a robotic arrayer. These were used to analyze interactions in six unique antigen mixtures ranging in concentration from 1.6 μg/mL to 1.6 ng/mL. The antigens were tagged with Cy3 and Cy5 fluorescent labels. Once the antigen-antibody binding reaction was complete, excess unbound antigens were washed off using phosphate-buffered saline and water at room temperature. Subsequently, after washing, the bound antigen-antibody interactions were detected using a microarray scanner at wavelengths of 532 and 635 nm. The authors found that such microarrays of antibodies could detect their corresponding antigens at concentrations as low as 1 ng/mL (Figure 12.15). In a complementary experiment, the authors generated six antigen arrays having 116 different antigens, which they probed with Cy3- or Cy5-labeled antibodies of varying concentrations. The antigen-antibody binding reaction was allowed to go to completion and excess unbound antibody was washed away using Phosphate buffer saline (PBS) and water at room temperature. The microarray slides were scanned at 532 and 635 nm. It was found that these antigen arrays allowed detection and quantitation of antibodies down to absolute concentrations of 100 pg/mL. These detection limits can further be improved by using high affinity and purity antibodies, thereby promising for high throughput and sensitive clinical applications.

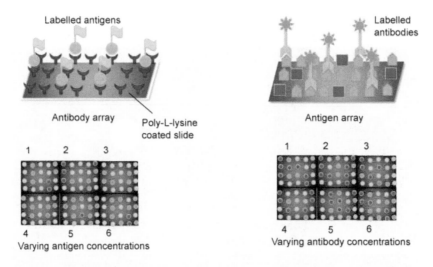

FIGURE 12.15   Application of antigen-antibody interactions. (From Haab et al., 2001.)

Steller and colleagues (Steller et al., 2005) amplified and subcloned 102 genes from N. meningitidis for expression in *E. coli* (Figure 12.16). Clones were grown for 16 hrs at 37°C in antibiotic-containing medium, following which protein expression was induced by the addition of isopropyl-β-D-thiogalactoside. The cells were harvested 4 hrs after induction and their proteins were purified based on specific Ni-NTA binding followed by analysis on SDS-PAGE. The 67 purified proteins obtained were then printed on nitrocellulose-coated glass slides using a robotic arrayer. The array was probed with sera from 20 convalescent patients by incubating it overnight at 4°C. The array was then washed with PBS and detection was carried out using Cy5-labeled secondary antibody. Excess detection antibody was washed off and the array was then dried and scanned. The researchers detected 47 immunogenic proteins, one of which showed a response in 11 of the patients.

Protein microarrays have been successfully used for the detection of several other disease biomarkers like cancer, autoimmune disorders as well as for diseases like Q-fever and other viral infections. Zhu with his group (Zhu et al., 2001) generated a yeast whole proteome array by expressing 5800 purified proteins on a single nickel-coated slide (Figure 12.17). The chips were probed with anti-GST antibodies to determine the reproducibility of protein immobilization. More than 93.5% of protein samples were found to give significant signals and over 90% contained 10 to 950 fg of

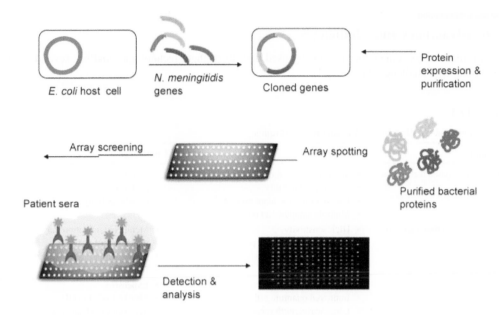

FIGURE 12.16　Application of biomarker detection. (From Steller et al., 2005.)

protein. To understand the potential applications of such whole proteome arrays, the authors screened the immobilized proteins for protein-protein and protein-lipid interactions. They used biotinylated calmodulin in the presence of calcium and phosphoinositide (PI) liposomes, respectively. Detection was carried out using Cy3-labeled streptavidin. Six known calmodulin targets and 33 potential partners were identified with 14 of these proteins possessing a consensus sequence. The PI liposomes identified 150 protein targets, of which 45 were membrane-associated and were predicted to have membrane-spanning regions. This study testified the tremendous potential of whole proteome arrays in identifying new protein targets.

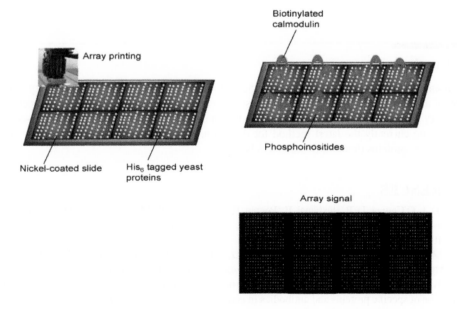

FIGURE 12.17　Application of protein interactions. (From Zhu et al., 2001.)

12.6 Advantages and Challenges

Protein microarrays come under the widely used high throughput technologies and has several advantages along with challenges (Table 12.1).

TABLE 12.1

Advantages and Challenges of Microarray Techniques

Technique	Advantages	Challenges
Fluorescence-based techniques	• Single antibody required • High reproducibility • Excellent for abundant proteins • Multiple samples can be assay	• Sensitivity low for abundant proteins • Cross-reactivity concerns
Single-walled carbon nanotubes	• High sensitivity • Multiplexed detection possible • Very less background signal • Photo bleach resistant	• Metal contamination may interfere with activity • Biological buffer insoluble • Hard to determine purity
Gold nanoparticles	• Good optical properties • Improved quantum efficiency • Large wavelength range • Photo bleach resistant	• Expensive • May be toxic to cells • Variations in shape and size of particles
Quantum dots	• Photostable • High fluorescent • Multicolor excitation possible	• Toxic • Mechanism not yet understood
Dye-doped silica nanoparticles	• High sensitivity • Biocompatible • Less dye leakage from particles • Particles do not aggregate • Photostable	• Difficult to synthesize
Bio-barcodes	• High sensitivity • Less detection time • Easily adapted to multiple protein targets	• Can only be used with known antibodies

12.7 Conclusions

Antigen-antibody interactions are highly specific and very strong. Apart from the simplicity of the production of antibodies, the antibodies themselves can act as antigens. This helps in easy detection using a secondary antibody such as fluorescent antibodies in case of microarrays. New SERS-based tags for antibodies are currently the latest method for detection. It is a highly sensitive method for detecting low-abundance proteins (femto-atto molar levels).

REFERENCES

Brennan, D. J., O'Connor, D. P., Rexhepaj, E., Ponten, F., & Gallagher, W. M. (2010). Antibody-based proteomics: Fast-tracking molecular diagnostics in oncology. *Nature Reviews Cancer, 10*(9), 605–617. https://doi.org/10.1038/nrc2902

Chan, S. M., Ermann, J., Su, L., Fathman, C. G., & Utz, P. J. (2004). Protein microarrays for multiplex analysis of signal transduction pathways. *Nature Medicine, 10*(12), 1390–1396. https://doi.org/10.1038/nm1139

Haab, B. B., Dunham, M. J., & Brown, P. O. (2001). Protein microarrays for highly parallel detection and quantitation of specific proteins and antibodies in complex solutions. *Genome Biology, 2*(2), research0004.1. https://doi.org/10.1186/gb-2001-2-2-research0004

LaBaer, J., & Ramachandran, N. (2005). Protein microarrays as tools for functional proteomics. *Current Opinion in Chemical Biology*, *9*(1), 14–19. https://doi.org/10.1016/j.cbpa.2004.12.006

MacBeath, G. (2002). Protein microarrays and proteomics. *Nature Genetics*, *32 Suppl*, 526–532. https://doi.org/10.1038/ng1037

Sanchez-Carbayo, M., Socci, N. D., Lozano, J. J., Haab, B. B., & Cordon-Cardo, C. (2006). Profiling bladder cancer using targeted antibody arrays. *American Journal of Pathology*, *168*(1), 93–103. https://doi.org/10.2353/ajpath.2006.050601

Steller, S., Angenendt, P., Cahill, D. J., Heuberger, S., Lehrach, H., & Kreutzberger, J. (2005). Bacterial protein microarrays for identification of new potential diagnostic markers for *Neisseria meningitidis* infections. *PROTEOMICS*, *5*(8), 2048–2055. https://doi.org/10.1002/pmic.200401097

Zhu, H., Bilgin, M., Bangham, R., Hall, D., Casamayor, A., Bertone, P., Lan, N., Jansen, R., Bidlingmaier, S., Houfek, T., Mitchell, T., Miller, P., Dean, R. A., Gerstein, M., & Snyder, M. (2001). Global analysis of protein activities using proteome chips. *Science*, *293*(5537), 2101–2105. https://doi.org/10.1126/science.1062191

Exercises 12.1

1. Which type of labeling method would you prefer not to use for an antigen which has only one epitope?
   a. Direct labeling
   b. Reverse-phase labeling
   c. Sandwich assay
   d. None of these

2. Which of the following particle type is most resistant to photo bleaching?
   a. Gold nanoparticles
   b. Raman dye-labeled particles
   c. Fluorescence-based dyes
   d. None of the above

3. Which of the following techniques is most cytotoxic?
   a. Gold nanoparticles
   b. Single-walled carbon nanotubes
   c. Fluorescent dyes
   d. Quantum dots

4. Which of the following is a label-based detection technique?
   a. Direct labeling
   b. Gold nanoparticles
   c. Surface plasmon resonance
   d. Single-walled carbon nanotubes

5. An array which can measure the levels of proteins in a given biospecimen is?
   a. Forward-phase protein array
   b. Antibody array
   c. Lipid array
   d. All of the above

6. Which of the following is *not* true for quantum dots?
   a. They are colloidal semiconductor nanocrystals
   b. They have stable light-emitting or scattering properties
   c. They emit excitons when light of lower band gap than that of the two semiconductors is incident
   d. They can be used as a novel method of labeling in protein arrays

7. A protein array with purified proteins printed on it belongs to which of the following categories of arrays when probed with serum samples containing autoantibodies?
   a. Forward-phase arrays
   b. Reverse-phase arrays
   c. Both of them
   d. None of them

8. Which of the following techniques couple with Raman spectrometry to detect spots using a Raman dye-labeled nanoparticle detection system?
   a. Electron microscopy
   b. Confocal microscopy
   c. Fiber optic microscopy
   d. Light microscopy

9. Which of the following components of a chemiluminescent detection system in protein arrays is enzyme linked?
   a. Substrate
   b. Primary antibody
   c. Printed proteins
   d. Secondary antibody

10. Which of the following is *not* true for direct labeling?
   a. Only single antibody is required
   b. High reproducibility
   c. High sensitivity in detecting low abundance proteins
   d. Multi-sample assay

11. Your aim is to identify novel serum biomarkers (autoantibodies) for developing an early diagnostic tool. You have a collection of 17000 cDNA expression clones and serum samples from 200 colon cancer patients and 100 healthy individuals. These expression clones carry the recombinant cDNA in a vector containing His-tag. What microarray approach would you take to identify novel autoantibodies in the serum?

Answers

1. c
2. a
3. d
4. a
5. b

6. c

7. b

8. c

9. d

10. c

11. The aim is to identify novel biomarkers. So, we do not have prior knowledge of the autoantibodies to be expected. In this scenario one can make a reverse-phase protein microarray (RPPM) and perform a sandwich assay. In this experiment, the cDNA expression clones are cultured and the recombinant proteins are expressed and purified using Ni-NTA. These purified His-tag proteins are spotted on microarray surface. These microarrays are treated with the serum samples obtained from colon cancer patients and healthy individuals. The antibodies attached to the antigens on the slide are detected with fluorescently tagged secondary antibodies. Based on the fluorescent signal intensity one can see if these antigens are differentially expressed in cancer patients and the healthy individuals. These differentially expressed antigens can be checked with the literature for their novelty.

13

Cell-Free Expression-Based Protein Microarrays

Preamble

Protein microarrays are miniaturized arrays containing small amounts of immobilized proteins, allowing the high-throughput study of a very large (100s–1000s) number of proteins. The biggest challenge in producing such arrays is the expression of proteins on such a large scale, without any loss in their structure and activity. Traditionally, recombinant protein expression has been carried out in systems such as *Escherichia coli*. Despite having various advantages of expressing recombinant expression in *E. coli*, there are some inherent problems associated with this expression system. Apart from being a very long and tedious process, the production, purification and maintenance of protein structure and functionality are difficult. The expressed proteins have a short shelf life, if not stored correctly, which makes protein storage difficult. One of the other challenges of using a prokaryotic system such as *E. coli* is that post-translational modifications may not be possible. The expression of eukaryotic proteins in a prokaryotic system such as *E. coli* often leads to the formation of inclusion bodies. These inclusion bodies are hard to purify and can be functionally inactive.

Cell-free expression involves the rapid, *in situ* synthesis of proteins from their corresponding DNA templates directly on the microarray surface. Protein arrays generated by this technique have shown great potential in eliminating the drawbacks of traditional cell-based methods. These cell-free protein expression systems use the basic mechanism of protein expression in cells, but without using intact live cells. The cell-free expression system should be able to utilize a wide variety of DNA templates and be able to express proteins with high reproducibility. The most important feature is to be able to produce proteins on-demand and avoid storage issues that lead to loss of activity of the produced protein. Detection and analysis of bound proteins should be simple. Although there are traditional methods for cell-free production of proteins, many novel microarray-based technologies for cell-free protein production have been developed, which will be discussed in this chapter.

Terminology

- **Cell-free expression:** Expression of proteins without using intact cells but using the translation machinery such as ribosomes, promoters, tRNAs, etc.
- **PISA:** Protein microarray in which a PCR product is used as the DNA template. *In vitro* transcription/translation (IVTT) mixture is added, and newly expressed protein is captured on the array using a suitable capture agent.
- **NAPPA:** Protein microarray in which the protein of interest is tagged (generally GST), expressed using IVTT, and expressed protein is captured on a slide coated with the anti-capture agent (anti-GST).
- **DAPA:** Protein microarray in which DNA template is attached to one slide, and another slide containing the capture agent is kept face-to-face with it. A permeable membrane containing the IVTT mixture is kept between the slides. Proteins are produced on the DNA slide, which is captured on the other slide.
- **MIST:** Protein microarray in which DNA is spotted on the array, while another mix containing the IVTT mixture is spotted precisely on the DNA spots. The newly expressed protein is captured on the array using a suitable capture agent.
- **HaloLink proteins:** The protein of interest is tagged with the HaloTag and the slide is coated with the Halo ligand. Thus, only the proteins containing the HaloTag are bound to the slide, and desired protein microarray is formed.

DOI: 10.1201/9781003098645-17

13.1 Commonly Used Techniques for Cell-Free Expression

Cell-free protein expression is the expression of proteins from template DNA, which is inserted in plasmids or present as purified PCR products. The addition of a crude cell lysate containing the required cellular machinery for protein production such as enzymes, ribosomes, tRNA, and exogenously added co-factors as nucleotides, Adenosine triphosphate (ATP), salts, essential amino acids are required for *in vitro* transcription and translation (IVTT) from the gene of interest. Additionally, based on cell-free expression machinery, such a system may also allow protein folding and post-translational modifications (PTMs) to produce structurally and functionally mature protein (Jackson, 2004; Katzen et al., 2005). There are different types of cell-free expression systems (Chandra & Srivastava, 2010) as shown in Figure 13.1.

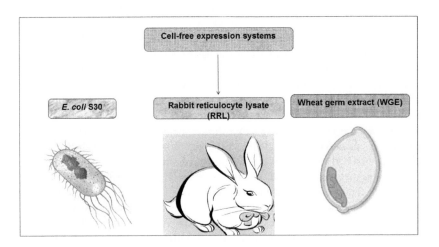

FIGURE 13.1 Different types of cell-free expression systems used for *in vitro* protein synthesis starting from DNA templates.

These systems contain all the necessary components and machinery for transcription and translation. Some factors like energy-generating components and essential amino acids need to be added to the system for successful protein synthesis. The following methods have been used for *in vitro* protein expression.

13.1.1 *Escherichia coli* Extract

The *Escherichia coli* extract (ECE) system has been in use for over the past 50 years. It consists of the crude cell-free extract containing ribosomes, tRNAs and translation machinery. *E. coli* S30 extracts are most commonly used (Figure 13.2). This is one of the best systems for the translation of protein from DNA templates as the source.

Actively growing and replicating *E. coli* cells can be used for extracting cell-free lysates. These cells that are in the process of growth and division are constantly producing proteins and other factors required for various cellular processes. Co-factors and enzymes such as RNA polymerase, peptidyl transferase are available in significant quantities due to cellular processes of transcription and translation taking place in the cell. The cells are lysed with a suitable buffer and then centrifuged at 30,000 ng to collect the supernatant containing the extract. Lysate that is extracted from such actively growing and dividing cells will contain all required cellular machinery to carry out *in vitro* protein synthesis and requires the addition of essential amino acids, nucleotides, salts and other energy-generating factors. However, it is incapable of carrying out post-translational modification (PTMs) of proteins due to the absence of required machinery for this process and often produces incomplete protein chains. The DNA templates obtained from bacterial sources are commonly used with cell-free lysate for IVTT of proteins.

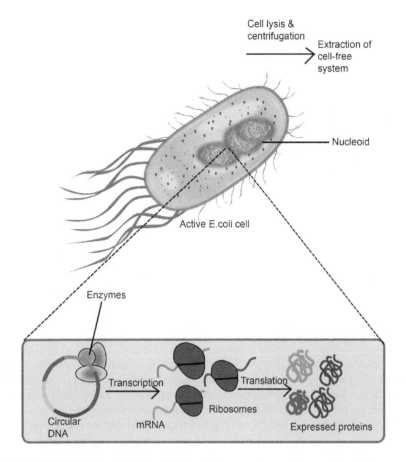

FIGURE 13.2 *Escherichia coli* S30 extract.

13.1.2 Wheat Germ Extract

The wheat germ extract (WGE) is also used for *in vitro* translation. The extract is from a plant embryo and contains all the machinery for protein synthesis such as initiation factors, translation factors and others. In addition to having all the components needed for protein expression, WGE has very low levels of endogenous mRNA, which aid in reducing the background translation by a considerable amount. WGE has been used for the efficient translation of exogenous RNA from a variety of different organisms.

One of the most commonly used eukaryotic cell-free expression systems is obtained from the embryo of the wheat seeds. The seeds are grinded and then sieved to remove their outer coating fragments. The embryos and other small particles are floated in an organic solvent such as cyclohexane. The floating embryos are quickly removed and dried to avoid any damage from the organic solvent. The dried embryos are then carefully sorted out such that only the good embryos without any endosperm coating are selected. The endosperm contains certain inhibitors of protein synthesis that must be removed. The selected embryos are washed thoroughly with cold water after which they are mixed with the extraction buffer and grinded. This solution must be centrifuged at 30,000 g at 4°C which results in the WGE forming a layer in between the top fatty layer fraction and the pellet at the bottom. This fraction can be separated and then purified by chromatographic methods to remove any components of the extraction buffer (Figure 13.3). This cell-free lysate is capable of synthesizing full-length eukaryotic proteins. This is a cell-free expression system that is capable of producing full-length proteins with correct folding and PTMs from bacterial, plant, or animal sources. Yields obtained in this system are however slightly lower than the *E. coli* and rabbit reticulocyte lysate (RRL).

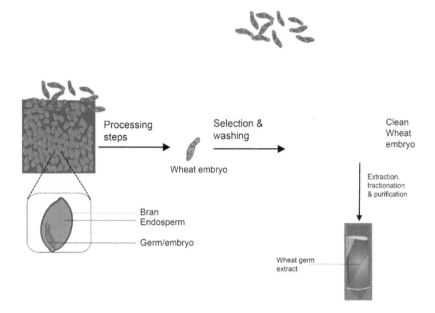

FIGURE 13.3 Wheat germ extract (WGE).

The DNA template is thawed and then placed on the ice during the preparatory process. For *in vitro* protein synthesis to take place, the DNA template must contain the gene coding for the protein of interest. In addition to this, there must be a promoter sequence, which can initiate the transcription process, a translation initiation sequence for binding of the ribosome as well as suitable termination sequences to correctly synthesize only the protein of interest. The thawed cell-free lysate containing the essential cellular machinery for protein synthesis is added to the DNA template followed by the other exogenous factors that are required for the process (Figure 13.4). All these are done while storing the template on ice to ensure that there is no loss of activity. The tube containing all the required components is then incubated at 30°C. Enzymes for transcription bind to the promoter sequences, and in the presence of other factors like ATP and nucleotides, they carry out the synthesis of the mRNA transcript. This mRNA is then translated into the corresponding protein with the help of ribosomes, tRNA, enzymes, and other factors required for the process.

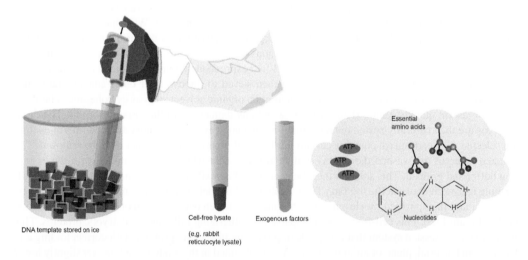

FIGURE 13.4 *In vitro* protein synthesis.

13.1.3 Rabbit Reticulocyte Lysate (RRL)

Another commonly used system for *in vitro* translation of proteins is the RRL. *In vivo*, the reticulocytes have the entire translation machinery as they produce hemoglobin. Thus, the reticulocyte lysate is ideal for the translation of proteins, as it contains all the requirements for protein expression such as initiation factors and ribosomes. Addition of micrococcal nuclease to the extract results in the degradation of the endogenous mRNA (Figure 13.5). RRL is more suitable for expressing of full-length eukaryotic proteins that require proper folding and PTMs.

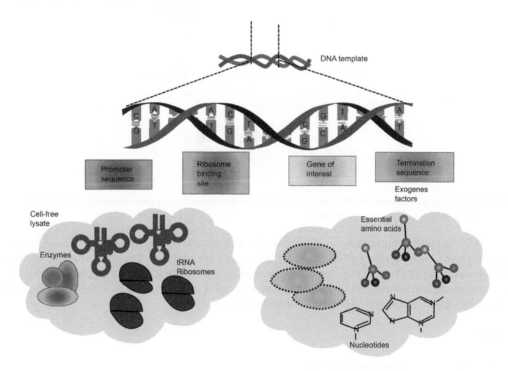

FIGURE 13.5 Rabbit reticulocyte lysate (RRL) is a mammalian cell-free system.

13.2 Cell-Free Expression-Based Protein Microarrays

Apart from the traditional methods for *in vitro* translation, new methods are being developed for cell-free protein expression. Many of these methods include the production of proteins on a solid substrate as on microarray. The following are some of the methods for *in situ* protein microarray production.

13.2.1 Protein *In Situ* Arrays (PISA)

PISA was the first technique of its kind to be developed. This technique was developed in 2001 which is also called the DiscernArray (He & Taussig, 2001). In this technique, a PCR product is used for translation. The PCR template has the sequence encoding the protein of interest, a T7 bacteriophage-derived promoter, a ribosome binding sequence for translation initiation, like a Shine-Dalgarno or Kozak sequence, an N- or C-terminal tag sequence, and termination sequences. This template is added as a free molecule onto the solid support – microtiter wells. The N- or C-terminal tag is used for immobilization of the expressed protein onto the array wells. This technique avoids DNA immobilization but is designed to immobilize the *de novo* synthesized protein. The DNA fragment is translated using any one technique like WGE or RRL, and protein is thus produced on the array. There are many methods to capture the bound proteins, e.g., the protein is engineered to contain a hexa-histidine tag, while the support pre-coated with the tag-capturing agent, nickel nitrilo-triacetic acid (Ni-NTA) is captured on the solid substrate. Thus, the target protein becomes captured on the microarray, while the non-specific, unbound materials can be removed by subsequent washing steps (Figure 13.6). Thus, a protein microarray of desired proteins can be produced.

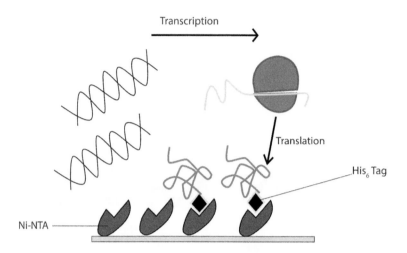

FIGURE 13.6   In the PISA method, PCR DNA encoding N- or C-terminal tag sequence expressed; the bound protein is specifically captured by the tag-capturing agent.

In PISA, the protein microarray surface is coated with a suitable tag-capturing agent that can immobilize the protein of interest through specific interactions once it is produced. The protein is expressed from its corresponding DNA using cell-free lysates such as *E. coli* S30 or RRL. The tagged protein is then captured on the array surface through the tag-capturing agent. PISA successfully overcame drawbacks of cell-based techniques such as protein insolubility and aggregation.

Merits

- Protein purification is not required.
- Rapid, single-step process.
- Specific protein attachment.
- Soluble proteins formed.

Demerits

- Possible loss of function during immobilization.
- Relatively high volume of cell-free lysate required.

13.2.2 Nucleic Acid Programmable Protein Array

This technique was developed by Ramachandran and group (Ramachandran et al., 2008). Nucleic acid programmable protein array (NAPPA) combines recombinant cloning technologies with cell-free protein expression. It replaces the cumbersome process of spotting the synthesized protein with the simpler process of spotting purified plasmid DNA. In this method, cDNA encoding a fusion of the protein of interest with a tag usually glutathione-S-transferase (GST) is expressed in plasmids by recombinant cloning.

- To prepare the microarray slides, a master mix, consisting of 4 components, namely plasmid-borne cDNA encoding a transcript for the protein of interest fused with the GST tag protein, bovine serum albumin (BSA) to improve the binding efficiency of cDNA, anti-GST antibodies and amine-amine cross-linker BS[3], is spotted onto the array chip.

- The array surface is coated with aminopropyltriethoxysilane (APTES), which helps in the binding of anti-GST antibodies cross-linked by BS[3]. BSA facilitates cDNA immobilization on the array surface.
- The array is then "activated" by adding RRL supplemented with T7 polymerase, RNase inhibitors and essential amino acids. Following transcription and translation of the ribosomal complex, the newly synthesized protein fused to the GST tag is then captured by the anti-tag (GST) antibodies attached to the array (Figure 13.7). Thus, a protein microarray is obtained.

This technique has been used to express proteins from over 10,000 unique cDNAs and has been used to express proteins from a variety of organisms. Furthermore, it is not confined to any particular class of proteins and has been efficiently (>90%) used to express various of proteins, including transcription factors, kinases and transmembrane proteins. Moreover, a wide range of sizes (from <50 to >100 kDa) can be efficiently expressed, making this a versatile tool for protein expression.

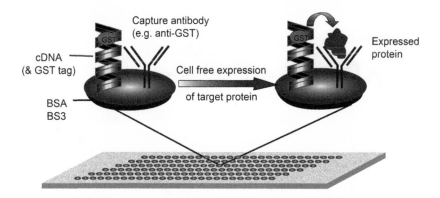

FIGURE 13.7 Protein microarray in which cDNA containing protein of interest is tagged (generally GST) and expressed using IVTT.

The expressed protein-containing GST tag is captured on a slide coated with the anti-capture agent (anti-GST antibody). An aminosilane-coated glass slide forms the array surface for NAPPA. To this, the NAPPA master mix is added which consists of BSA, BS[3], GST-tagged cDNA and anti-GST capture antibodies. The BSA improves the efficiency of immobilization of the cDNA onto the array surface while the BS[3] cross-linker facilitates the binding of the capture antibody. The cDNA is expressed using a cell-free extract to give the corresponding protein with its GST tag fused to it. This tag enables the capture of the protein onto the slide using anti-GST antibodies. NAPPA technique can generate very high-density arrays but the protein remains co-localized with cDNA.

Merits

- No need to express and purify protein separately.
- Expression in the mammalian milieu (natural folding).
- Proteins produced just-in-time for assay.
- Shelf life not an issue.
- Access to all cloned cDNAs.
- Express and capture more than protein spotting arrays.
- Retains functionality of traditional protein arrays.
- Arrays stable on the bench until activated.

Demerits

- Cloning procedure required.
- Pure protein array not produced.
- Peptide tags may lead to sterical effects blocking important binding domains.
- Functionality of proteins.

13.2.3 Multiple Spotting Technique

Another method of making a protein array is the multiple spotting technique (MIST), which was developed by group of researchers (Angenendt et al., 2006). Here, the support such as a slide is pre-coated with a protein capture agent. The first spot printed on the slide consists of the DNA template. The second spot printed contains the *in vitro* translation mixture, which is added exactly on top of the spot containing the DNA template (Figure 13.8). The translation of the proteins takes place, and the protein-capture agent captures the newly formed protein on the slide, and an array is constructed.

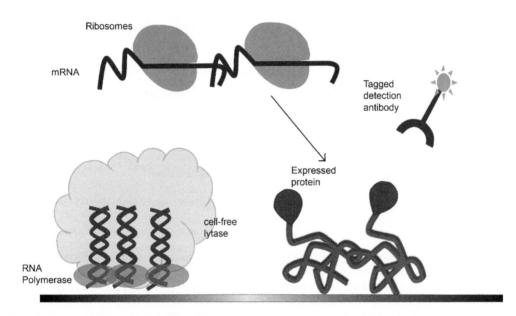

FIGURE 13.8 Multiple spotting technique (MIST).

The first spotting step of the MIST, which is also capable of producing high-density arrays, involves the addition of template DNA onto the solid array support. The template DNA can even be in the form of an unpurified PCR product, one of the major advantages of this technique. The second spotting step involves the addition of the cell-free lysate directly on top of the first spot. Transcription and translation can begin only after the second spotting step. The protein expressed from the template DNA binds to the array surface by means of non-specific interactions, one of the drawbacks of this procedure. A detection antibody specific to the protein of interest is then added which indicates protein expression levels through a suitable fluorophore.

Merits

- Unpurified DNA products used as a template.
- Very high-density protein arrays generated.

Demerits

- Loss of signal intensity with prolonged incubation time.
- Non-specific protein binding.
- Time-consuming process.

13.2.4 DNA Array to Protein Array

This concept was developed by He et al. (2008). In the DNA array to protein array (DAPA), two different slides are used. One of the slides coated with Ni-NTA has the PCR-amplified DNA fragments, which encode the protein of interest fused with a tag immobilized onto it; while the other Ni-NTA slide has the protein tag-capturing agent. The slides are placed face-to-face and a permeable membrane is kept between them. The permeable membrane has cell-free lysate, and thus protein expression is initiated in between the two slides. The newly synthesized proteins are produced on the slide with the DNA template, which then penetrates the membrane and becomes immobilized on the surface of the slide bearing the tag-capture reagent. Thus, a replica of the DNA array is formed on the captured slide (Figure 13.9). DAPA leads to the construction of a pure protein array, i.e., an array with no DNA contamination, as the two are kept separate throughout the experiment. Another important advantage of DAPA is that the DNA template slide is reusable, i.e., the proteins can be printed many times from the same DNA, making the process less time-consuming and cost-effective.

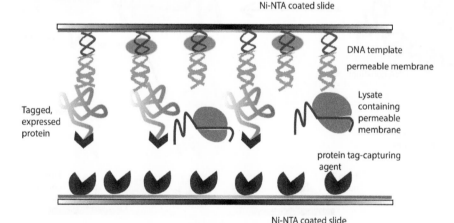

FIGURE 13.9 The microarray slide surface is coated with nickel-nitrilotriacetic acid (Ni-NTA), which acts as a useful capture agent.

A permeable membrane containing lysate is inserted between the slides holding the DNA template and the protein tag-capturing agent. Through its capture agent, the produced protein slowly permeates the membrane and becomes fixed on the slide surface. This technique allows for several reuses of the DNA template array. On a Ni-NTA-coated slide, PCR-amplified DNA that codes for the desired protein is immobilized. Between a slide with the protein tag-capturing agent and a slide with the immobilized DNA template is a permeable membrane that has been soaked in the cell-free extract. Invading the membrane, the newly expressed proteins attach to the protein purification slide.

Merits

- Reusable DNA template array.
- Pure protein array generated.
- DNA template array can be stored for long durations.

Demerits

- Broadening of spots due to diffusion.
- Not ascertained if multimeric proteins assemble effectively.
- Time-consuming process.

13.2.5 HaloLink Protein Array

This amalgamation of technologies developed by Promega is useful in generating tightly immobilized arrays. The HaloLink array consists of a DNA construct encoding the gene of interest fused with the "HaloTag", a 33 kDa mutated bacterial hydrolase (Figure 13.10). The protein is constructed using a cell-free expression system (WGE or RRL). The newly formed proteins are captured on a polyethylene-glycol-coated glass slide, which has been activated by HaloTag ligands. The HaloTag fused to the protein of interest binds the Halo ligand on the slide by covalent bonding, thus enabling the capture of the desired protein. It also allows oriented capture of proteins, hence keeping the protein activity unaffected. The mode of interaction between HaloTag and its ligand is through covalent bonding, thereby ensuring firm capture of the protein on the array surface without any material loss during washing.

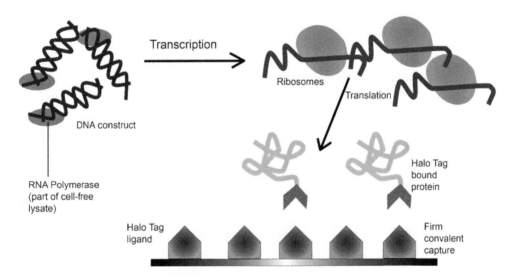

FIGURE 13.10 HaloTag technique for microarray.

The slide is activated with the HaloTag ligand, which captures the expressed protein through firm covalent interactions thereby preventing any material loss and ensuring oriented capture of the protein. The HaloTag-fused protein is expressed using lysates like RRL or WGE and covalently captured onto the array surface through the HaloTag ligand. The specific interaction ensures oriented capture of the protein thereby preventing any possible functional loss.

Merits

- Strong covalent bond between protein and ligand.
- No material loss during washing.
- Oriented capture of protein.
- No non-specific adsorption.
- Easy quantification.
- No need for a microarrayer printer.

Demerits

- Possible loss of function on binding to HaloTag.
- HT application will require optimization of printing.

13.3 Applications

The microarrays spotted with proteins expressed using a cell-free expression system have a variety of applications as given below (Figure 13.11)

- Biomarker discovery.
- Immunogenicity studies.
- Protein-protein interaction studies.
- PTM studies.
- Simultaneous screening of a large number of proteins.

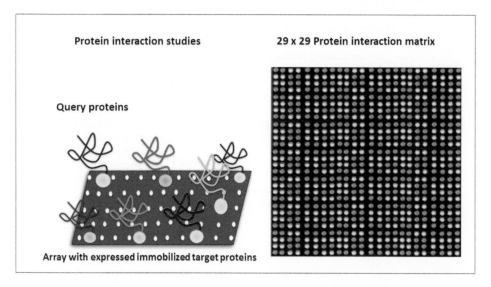

FIGURE 13.11 Application of cell-free expression system for protein-protein interaction study.

Ramachandran and group tested the use of NAPPA microarrays by immobilizing 29 sequence-verified human genes involved in replication initiation on the array surface and then expressing them in duplicate with RRL (Ramachandran et al., 2004). The expressed proteins bound to the anti-GST antibodies present on the array surface. The authors used each of these expressed proteins to probe another duplicate array of the same 29 proteins thereby generating a 29 × 29 protein interaction matrix. 110 interactions were detected between proteins of the replication initiation complex, of which 63 were previously undetected.

13.4 Advantages and Challenges

There are several advantages and challenges of Microarray techniques as given in Table 13.1.

TABLE 13.1

Advantages and Challenges of Microarray Techniques

Technique	Advantages	Challenges
PISA	• Creates soluble proteins *in situ*. • Overcomes common problems (protein insolubility, degradation, and aggregation issues) endemic during protein expression in prokaryotic systems. • Only tagged protein remains on the array.	• Requires immediate utilization of PCR-produced DNA. • Not cost-effective as a relatively large volume of cell-free lysate is required. • Hexa-histidine tag may interfere with proper protein folding.
NAPPA	• Use of mammalian expression systems allows efficient folding. • Access to a wide variety of cloned cDNAs. • Shelf life not an issue: cDNA arrays stable for long periods at RT, and more difficult-to-store proteins are produced just before assay. • Cost-effective: low required volume of cell-free lysate. • Effective process: over 95% of proteins tested express and capture well. Discrete spots are obtained.	• Need for time-consuming cloning before generating array, or alternately dependent on available clones. • Need to clone the gene of interest as its GST fusion. • Pure protein arrays not produced: expressed protein remains co-localized to cDNA. • Peptide tags may produce steric hindrance while studying protein interactions. • Correct functionality of proteins always remains in doubt during cell-free expression.
MIST	• Non-purified PCR product can be used as DNA template. • Very high-density protein arrays generated.	• Non-specific protein binding can occur. • Time-consuming. • Loss of signal intensity occurs with prolonged incubation.
DAPA	• A pure protein array, free of DNA is generated. • Allows the generation of multiple protein arrays from a single DNA template. • The DNA template array can be stored for long periods.	• The broadening of spots may occur due to protein diffusion. • Multimeric proteins may not assemble efficiently. • Time-consuming process.
HaloLink protein arrays	• Covalent bond allows firm immobilization of proteins. • Proteins are captured in an oriented manner with no non-specific adsorption. • Little functional or quantitative losses of materials during washing steps. • Accurate quantification of protein possible. • No requirement for microarray printer as gaskets for printing are provided.	• Has not been validated for high density-large protein number arrays, although theoretically possible. • Loss of protein function may occur due to binding to the HaloTag.

13.5 Conclusions

The usage of protein microarrays has made it possible to screen a large number of proteins simultaneously, leading to high-throughput studies. Many types of cell-free expression-based microarrays have shown promising results (Figure 13.12). NAPPA and DAPA aid in building pure protein arrays with no DNA contamination. Proteins have relatively shorter shelf lives and protein microarrays help in increasing the shelf life of the proteins. Making protein arrays is also cost-effective, as very small amounts of reagents are used. Moreover, the DNA templates are reusable, which reduces the cost further. The production of proteins becomes much easier, and storing DNA arrays helps in reproducing the arrays when required. Microarrays have tremendous potential in clinical applications as well as non-clinical research such as detection of protein-protein interaction and biological screening.

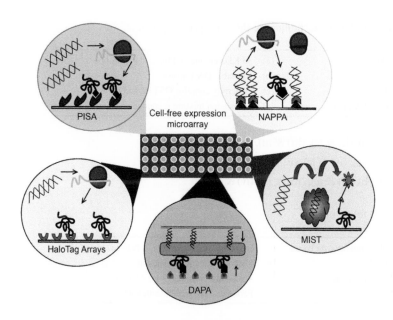

FIGURE 13.12 An overview of cell-free expression-based microarrays.

REFERENCES

Angenendt, P., Kreutzberger, J., Glökler, J., & Hoheisel, J. D. (2006). Generation of high density protein microarrays by cell-free in situ expression of unpurified PCR products. *Molecular & Cellular Proteomics, 5*(9), 1658–1666. https://doi.org/10.1074/mcp.T600024-MCP200

Chandra, H., & Srivastava, S. (2010). Cell-free synthesis-based protein microarrays and their applications. *PROTEOMICS, 10*(4), 717–730. https://doi.org/10.1002/pmic.200900462

He, M., Stoevesandt, O., Palmer, E. A., Khan, F., Ericsson, O., & Taussig, M. J. (2008). Printing protein arrays from DNA arrays. *Nature Methods, 5*(2), 175–177. https://doi.org/10.1038/nmeth.1178

He, M., & Taussig, M. J. (2001). Single step generation of protein arrays from DNA by cell-free expression and in situ immobilisation (PISA method). *Nucleic Acids Research, 29*(15), 73e–773. https://doi.org/10.1093/nar/29.15.e73

Jackson, A. M. (2004). Cell-free protein synthesis for proteomics. *Briefings in Functional Genomics and Proteomics, 2*(4), 308–319. https://doi.org/10.1093/bfgp/2.4.308

Katzen, F., Chang, G., & Kudlicki, W. (2005). The past, present and future of cell-free protein synthesis. *Trends in Biotechnology, 23*(3), 150–156. https://doi.org/10.1016/j.tibtech.2005.01.003

Ramachandran, N., Hainsworth, E., Bhullar, B., Eisenstein, S., Rosen, B., Lau, A. Y., Walter, J. C., & LaBaer, J. (2004). Self-assembling protein microarrays. *Science, 305*(5680), 86–90. https://doi.org/10.1126/science.1097639

Ramachandran, N., Raphael, J. V., Hainsworth, E., Demirkan, G., Fuentes, M. G., Rolfs, A., Hu, Y., & LaBaer, J. (2008). Next-generation high-density self-assembling functional protein arrays. *Nature Methods, 5*(6), 535–538. https://doi.org/10.1038/nmeth.1210

Exercises 13.1

1. Which is the least suitable method for protein expression of a protein that has a high amount of post-translational modifications?

    a. *E. coli* expression

    b. Wheat germ extract

    c. Rabbit reticulocyte lysate

    d. None of these

2. Match the following techniques to the appropriate option:

(i)	DAPA	(a)	33 KDa engineered bacterial hydrolase
(ii)	NAPPA	(b)	Reusable DNA arrays
(iii)	MIST	(c)	Generally contains GST tagged proteins
(iv)	HaloLink protein assay	(d)	Translation mixture is spotted on top of DNA spot

1. (i) & (b), (ii) & (c), (iii) & (d), (iv) & (a)
2. (iii) & (a), (iv) & (c), (ii) & (d), (i) & (b)
3. (i) & (b), (ii) & (c), (iii) & (a), (iv) & (d),

3. Match the following:

(i)	Rabbit reticulocyte lysate	(a)	Formation of covalent bonds between ligand and tag
(ii)	DAPA	(b)	Micrococcal nuclease added
(iii)	HaloLink proteins	(c)	Array surface coated with APTES
(iv)	NAPPA	(d)	Permeable membrane with RRL

4. What is the function of the BS^3 linker?
   a. It helps in attachment of the template DNA to slides
   b. It helps in cross-linking of proteins
   c. It helps in attachment of the Halo ligands to the slides
   d. It helps to remove non-specific proteins from the sample

5. If a PCR DNA encodes a Hexa-His tag, what should an ideal capturing agent be for the translated protein in a PISA-based protein microarray experiment?
   a. Streptavidin
   b. Biotin
   c. Ni-NTA
   d. Anti-GST

6. Which of the following attribute(s) makes HaloTag a better technology than traditional tags in protein arrays?
   a. The HaloTag is a modified haloalkane protein tag with a small molecular weight
   b. The HaloTag forms a covalent association with its synthetic ligand unlike other tags
   c. It enables rigorous washing steps leading to less background
   d. All of the above

7. While fabricating a multiple spotting technique (MIST)-based protein array, a cell-free expression system is directly added to the DNA template spot in which spotting step?
   a. First spotting step
   b. Second spotting step
   c. Third spotting step
   d. Fourth spotting step

8. Which of the following is *not* an example of a printing parameter?
   a. Temperature
   b. Humidity
   c. Light
   d. Pin washing

9. An ideal cell-free expression system for mammalian proteins is?
   a. Wheat germ extract
   b. *E. coli* cell lysate
   c. Human Hela cell lysate
   d. None of the above

10. Which of the following component is *not* required for a successful cell-free protein synthesis reaction?
    a. DNA template
    b. Promoter
    c. Proteases
    d. Translation initiation sequence

11. You want to quantify a protein using protein microarrays but you do not have access to a microarray printer. In this scenario, which type of protein microarrays could you use? Also, the loss of material during the washing steps is an issue in most of the microarray types. How can one address this problem?

Answers

1. a
2. (ii) & (c), (iv) & (a), (i) & (b), (iii) & (d)
3. (iii) & (a), (iv) & (c), (ii) & (d), (i) & (b)
4. b
5. c
6. d
7. b
8. c
9. c
10. c
11. Based on the conditions mentioned above, one can use HaloTag microarrays. HaloTag microarrays are known to be used for accurate quantification of proteins. For making HaloTag protein microarrays, one does not require microarray printer. There are kits, which are provided with gaskets, which help in printing the protein spots on the microarray surface. The proteins spotted on the surface are covalently attached to the microarray surface and this reduces the material loss during the vigorous washing steps.

14

Nucleic Acid Programmable Protein Arrays

Preamble

Various important cell processes such as ligand-receptor reaction, hormonal activity, enzymatic catalysis, DNA replication, respiration, and growth occur because of the protein-protein interactions. To understand cellular processes, studying the protein-protein interactions is most important. This involves the production of desired proteins and studying them *in vitro*, which is highly time-consuming and demanding process. Since protein production takes a lot of time, very few proteins can be studied at a time. It is also difficult to maintain the activity of the protein produced. Moreover, the shelf life of proteins is very low. Hence, there is a need to develop new methods for the production of better-quality proteins in a high-throughput manner.

One of the most revolutionizing technologies developed in recent times is protein microarrays. A protein microarray has solid support, on which thousands of proteins can be spotted and studied at once. This has become possible due to the advancements of cell-free expression systems. The construction of protein arrays does not take a lot of reagents and therefore is cost-effective and a large number of proteins can be screened at once. One of the cell-free expression-based techniques developed for the construction of protein microarrays is nucleic acid programmable protein arrays (NAPPA), which is discussed in this chapter.

Terminology and Basics

- **APTES (aminopropyltriethoxysilane) coating:** APTES is used for coating of the glass slide so that it becomes amenable to attach DNA and proteins.
- **GST tag:** Tag fused at the C-terminal end of the protein, for capture on anti-glutathione S-transferase (GST) capture antibodies.
- **BS³ linker:** A water-soluble, non-cleavable, linker with a spacer of eight Carbon atoms between two hydroxysuccinimide groups which links the primary amine groups to help tether the antibody to the glass slide.
- **BSA:** The bovine serum albumin (BSA) enhances DNA binding in NAPPA chemistry.
- **Recombinational cloning:** A novel site-specific recombination technique for transferring DNA sequences, which allows one universal strategy to move DNA sequence to any vector. Recombination cloning is a promising approach for high-throughput genomics and proteomics applications.
- **Site-specific recombination:** A genetic recombination technique where DNA strand exchange takes place between regions possessing a reasonable degree of sequence homology. Specific recombinase enzymes cleave the DNA backbone and interchange DNA helices between specific sites on two different molecules. The common site-specific recombination technologies currently in use are the Gateway Technology (Invitrogen) and the Creator Technology (BD Clontech).
- **BP reaction:** A site-specific recombination reaction between the attB and attP sites leads to the generation of the master or entry clones, which can be used at any time for specific purposes (Figure 14.1).
- **LR reaction:** The main reaction pathway of the Gateway system consisting of a recombination reaction between a master clone and a destination vector used to generate the expression clones (Figure 14.1). These expression clones can be used for a variety of applications.

DOI: 10.1201/9781003098645-18

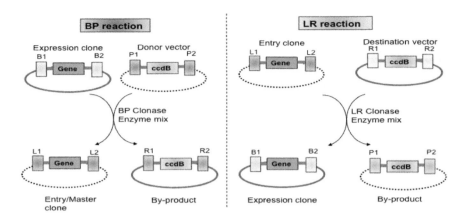

FIGURE 14.1   Recombinational cloning steps: BP and LR reactions.

Gateway cloning: A powerful new recombinational cloning technology that facilitates protein expression and cloning of PCR products by using site-specific recombination enzymes rather than restriction endonucleases and ligases (Park & LaBaer, 2006). This technique makes use of a master clone having a particular gene that can be rapidly transferred to desired destination vectors and thereby provides a significant benefit over conventional cloning (Figures 14.2 and 14.3).

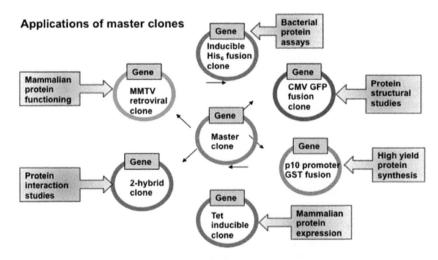

FIGURE 14.2   Master clones generated by recombinational cloning can be used for several applications.

The BP reaction of Gateway cloning is a site-specific recombination reaction between the attB site of an expression clone or a PCR product and the attP site of a donor vector in the presence of the BP Clonase enzyme master mix. The reaction is incubated for just an hour at 25°C to obtain the entry or master clones containing the gene of interest. Once this master clone, flanked by attL sites, is produced, it can then be transferred into any destination vector to produce expression clones for a specific desired application. The reaction yields more than 90% correct clones. The LR reaction is essentially the reverse of the BP reactions where the master clone, flanked by attL sites, recombines with a destination vector with attR sites. This reaction takes place in the presence of the LR Clonase enzyme mix and results in the transfer of the gene from the master clone to the destination vector to produce an expression clone for a specific purpose. This reaction enables the generation of several expression clones for various applications in a very short time, thereby providing a significant advantage over conventional cloning techniques.

The gene in the master clone can be transferred to various destination vectors by means of the LR reaction to produce expression clones for several applications. Proteins can be efficiently expressed in bacterial, yeast, and mammalian systems and used for a variety of applications such as structural and

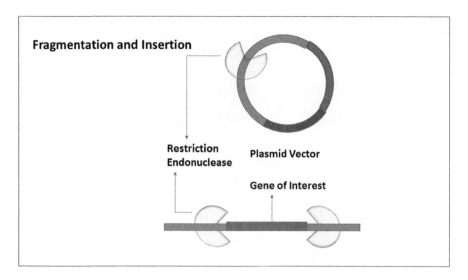

FIGURE 14.3 Recombinational cloning.

functional studies, protein interaction studies, protein assays, producing high yields of proteins for experimentation, etc. The rapid recombination between clones that is possible with the GATEWAY system cannot be done with conventional cloning techniques, due to which the GATEWAY protocol is now being extensively adopted.

14.1 NAPPA – Nucleic Acid Programmable Protein Array

Protein microarrays consist of solid surfaces on which the proteins of interest are spotted. Earlier, each protein was spotted on the slide surface using chemical techniques. This required manual production of each protein, which was very tedious. The self-assembling protein microarray or NAPPA approach introduced by Ramachandran (2004) used spotting of expression plasmids containing cDNAs of interest on the array surface and expression of proteins *in situ* by a mammalian cell-free expression system at the time of assay. All proteins were expressed with fusion tags, which correspond to capture agents printed along with the plasmid DNA and used to capture the protein immediately after translation. By producing the proteins just-in-time for assay, the opportunity for its denaturation is significantly reduced, and the use of a mammalian transcription/translation system encouraged natural protein folding for mammalian protein. The NAPPA microarray is a highly innovative cell-free expression-based technology. It helps in the production of thousands of proteins simultaneously and also captures them on the slide to form the array as and when they are formed.

In NAPPA, cDNA (containing a tag, usually glutathione S-transferase [GST]) of the desired proteins are spotted along-with BSA, BS[3] and capture antibody (anti-GST antibody) on the functionalized slide (Figure 14.4). For activation of the array, a cell-free expression mixture containing *in vitro* transcription and translation (IVTT) mix (rabbit reticulocyte lysate (RRL), T7 polymerase, amino acid mixtures,

Printing master mix

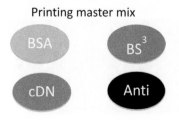

FIGURE 14.4 Master mix for NAPPA printing consists of cDNA with GST tag, BS[3] cross-linker, BSA, and capture antibody (anti-GST).

RNase inhibitor, etc.) is added onto the slide. This leads to the production of desired proteins from the cDNA. The protein produced contains a tag, which binds to the capture agent coated on the slide. Thus, a protein replica is formed in place of the DNA array (Figure 14.5).

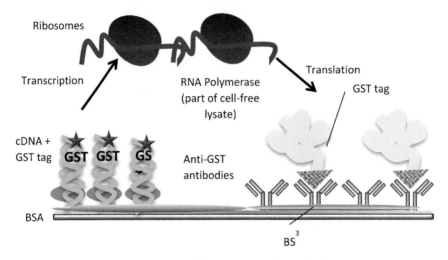

FIGURE 14.5 Working principle of nucleic acid programmable protein arrays.

The master mix containing cDNA (with GST tag), BSA, BS^3, and anti-GST antibodies are printed on the array surface. Adding RRL to the arrays carries out protein expression. Antibodies capture the newly synthesized proteins through the GST tag, producing protein microarray. An aminosilane-coated glass slide forms the array surface for NAPPA. To this, the NAPPA master mix is added which consists of BSA, BS^3, GST-tagged cDNA, and anti-GST capture antibodies. The BSA improves the efficiency of immobilization of the cDNA onto the array surface, while the BS^3 cross-linker facilitates the binding of the capture antibody. The cDNA is expressed using a cell-free extract to give the corresponding protein with its GST tag fused. This tag enables the capture of the protein onto the slide by means of anti-GST antibodies (Figure 14.6). NAPPA technique can generate very high-density arrays but the protein remains co-localized with cDNA.

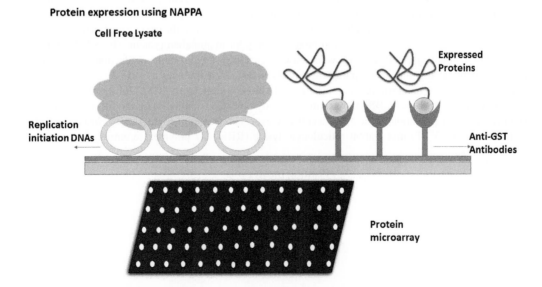

FIGURE 14.6 Nucleic acid programmable protein array (NAPPA).

14.2 Workflow of NAPPA

To ensure the construction of the NAPPA protein microarray, one must carefully design each component. The design and role of each component of the NAPPA workflow are as follows.

14.2.1 Preparation of Master Mix for Printing

The solid support for printing the array is generally a glass slide. But other supports such as gold, nitrocellulose, and hydrogel also may be used. Depending upon the surface of the microarray, the surface chemistry for printing cDNA differs. The glass slide is coated with aminopropyltriethoxysilane (APTES) or aminosilane and it contains a large amount of positively charged amino groups, which bind to the negatively charged phosphate groups on DNA (Figure 14.7). The slides are exposed to UV or baked at 85°C enabling the strong covalent attachment of DNA on the silane-coated slides.

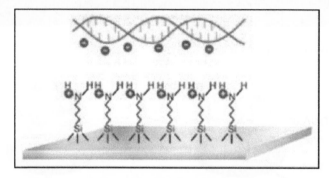

FIGURE 14.7 Aminosilane coating of the glass slide. The positive groups of amine on the slide surface covalently bind to the negatively charged phosphate groups of DNA, which enables the immobilization of DNA molecules onto the slide surface.

If the cDNA is biotinylated (using psoralen-biotin), the slide is coated with avidin along with aminosilane. This leads to a strong biotin-avidin binding enabling the DNA immobilization onto the slide surface. For protein capture, a BS^3 cross-linker and anti-tag antibody are also spotted on the silane-coated glass slide. This linker helps in the formation of amine-amine bonds and thus is used for tethering the capture antibody on the glass slide. The cDNA for the protein of interest is designed for two purposes: (1) the cDNA should be captured on the slide before protein production, and (2) it must have a tag, which helps in the identification and capture of the expressed protein. The cDNA plasmids are designed to form a fusion tag at either the N- or C-terminal. Generally, plasmids are designed to add a C-terminal GST tag. BSA is added to this master mix as it enhances the binding of the DNA to the slide. The master mix is printed on the slide using manual spotting or using an automated microarray.

14.2.2 Protein Production and Capture

For protein production using IVTT system, a cell-free expression mixture containing rabbit reticulocyte lysate, T7 polymerase, amino acid mixtures and RNase inhibitor is added onto the slide. The proteins expressed from the cDNA contain a tag (usually GST tag) at the C-terminal of the protein. The anti-GST antibody is coated on the aminosilane slide, with the BS^3 cross-linker. The synthesized protein is immobilized to the slide by antigen-antibody reactions forming a protein array. PicoGreen dye is used to test DNA printing quality, and anti-GST antibody is used to test protein expression. The protein-specific antibody is used to test protein-specific expression (Figure 14.8). Similar principle forms the basis of NAPPA (Ramachandran et al., 2004) (Figure 14.9).

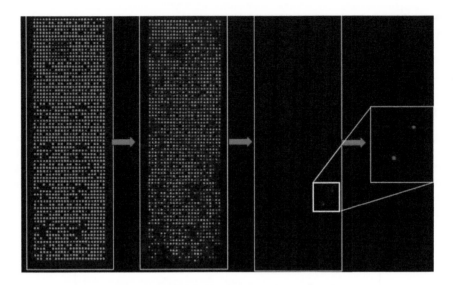

FIGURE 14.8   Quality control of NAPPA array.

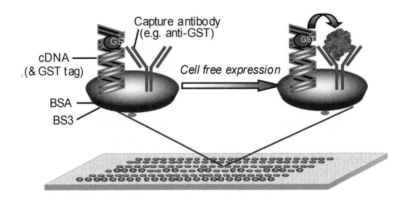

FIGURE 14.9   Schematic for NAPPA protocol.

14.3 Applications of NAPPA

14.3.1 Autoantibody Biomarker Screening

Antibodies to tumor antigens have advantages over other serum proteins as potential cancer biomarkers because they are stable, highly specific, easy to purify from serum, and are readily detected with well-validated secondary reagents. These antibodies directed at self-antigens are referred to as autoantibodies. NAPPA arrays have been used to identify autoantibody biomarkers in sera that can be readily used for the early detection of cancers. For autoantibody screening, samples such as serum or cell lysate are added to the chip. If target antigens are present in the sample, autoantibodies bind to their targets, and this can be detected using a labeled anti-IgG antibody (Anderson et al., 2008) (Figure 14.10).

14.3.2 Protein Interactions

One key application of NAPPA is to test protein-protein interactions (Figure 14.11). Typically, this is done by probing an array of proteins with a purified query protein. NAPPA can also employ co-expression of the target and the query protein by transcribing and translating them in the same extract. To do this, simply appropriate query DNA was added into the IVTT before applying it to the array. For fluorescent detection,

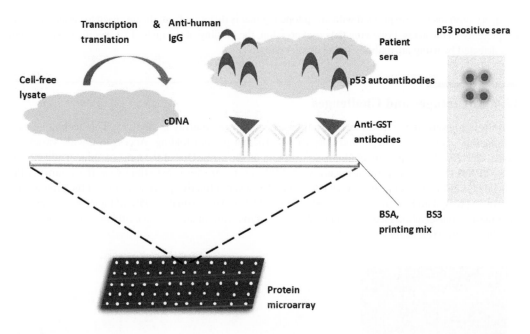

FIGURE 14.10 Detection of p53 autoantibodies using NAPPA microarrays. The authors generated protein microarrays based on NAPPA expression, which they probed with diluted sera of breast cancer patients having p53 autoantibodies. Detection was carried out by means of HRP (horseradish peroxidase) linked anti-human IgG. This study detected p53 autoantibodies by means of NAPPA microarrays, which was confirmed by ELISA. The p53 levels were found to be directly related to tumor burden with serum antibody concentration decreasing after neoadjuvant chemotherapy (Anderson et al., 2008).

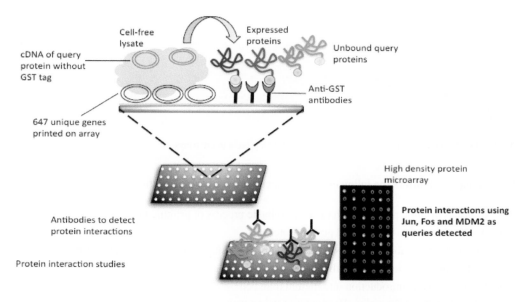

FIGURE 14.11 Protein-protein interaction study using NAPPA arrays. Ramachandran et al. (2004) tested the use of NAPPA microarrays by immobilizing 29 sequence-verified human genes involved in replication initiation on the array surface and then expressing them in duplicate with RRL. The expressed proteins are bound to the anti-GST antibodies on the array surface. The authors made use of each of these expressed proteins to probe another duplicate array of the same 29 proteins thereby generating a 29 × 29 protein interaction matrix. One hundred ten interactions were detected between proteins of the replication initiation complex, of which 63 were previously undetected.

the query proteins were expressed with an epitope tag that is different from the one used to capture the target proteins on the array. Following the co-expression, the binding of the query to a specific target protein was detected by using an antibody to the query epitope.

14.4 Advantages and Challenges

NAPPA is a promising cell-free expression-based protein microarray technology. This technique uses mammalian expression systems and thus allows efficient protein folding. Access to a wide variety of cloned cDNAs, allows spotting of almost any protein on the array. One of the most prominent advantages of NAPPA is the long shelf life of the arrays. NAPPA arrays are very cost-effective as the volume of the cell-free lysate and the template is very low. NAPPA is a very effective process as 95% of the proteins are usually expressed and captured on the slide (Figure 14.12). Approximately 95% of the protein expressed are kinases, transcription factors, and membrane proteins without any bias toward low or high molecular weight proteins (Ramachandran et al., 2008).

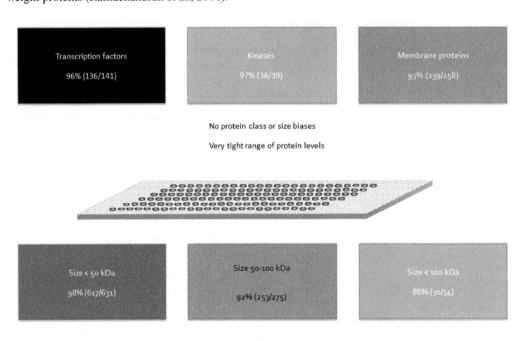

FIGURE 14.12 Success of protein expression using NAPPA protein microarrays.

NAPPA approach offers the following advantages over traditional methods:

- Replaces printing proteins with the more reliable process of printing DNA.
- Avoids the need to express and purify proteins.
- Avoids concerns about protein shelf life because the proteins are made fresh at the time of assay.
- Protein integrity: uses mammalian machinery to synthesize proteins.
- No tedious gene introduction into mammalian cells.
- Interaction is not limited to the context of the nucleus.
- Gene toxicity and auto-activation of reporter genes are not an issue.
- Expensive mass spectrometers are not required.
- Analysis of multimeric complexes.
- Post-translational modifications.

- Protein production and obtaining data in real-time.
- Sequence-verified plasmid templates provide good quality controls.
- Attachment/detection scheme changeable.

However, like any other technology, NAPPA has certain limitations. The process is time-consuming, as cloning of each cDNA template needs to be done before generating an array, or it depends on the availability of clones. Each cDNA needs to be cloned with fusion tags (e.g., GST). Another problem associated with NAPPA is that producing pure protein arrays is not possible, as the expressed protein remains co-localized with cDNA. The peptide tags may produce steric hindrance while studying protein interactions. Finally, it is difficult to assess the functionality of expressed proteins after IVTT steps.

14.5 Conclusions

NAPPA is a very effective technique for the production of protein microarrays. It has already been used in various studies for immunological screening, screening of biomarkers and studying protein-protein interaction. The arrays can be stored for long, as NAPPA uses DNA for printing, and proteins can be produced on demand. Integrating NAPPA microarrays with label-free detection techniques such as surface plasmon resonance imaging would be an interesting future direction for several high-throughput proteomic applications.

REFERENCES

Anderson, K. S., Ramachandran, N., Wong, J., Raphael, J. V., Hainsworth, E., Demirkan, G., Cramer, D., Aronzon, D., Hodi, F. S., Harris, L., Logvinenko, T., & LaBaer, J. (2008). Application of protein microarrays for multiplexed detection of antibodies to tumor antigens in breast cancer. *Journal of Proteome Research*, *7*(4), 1490–1499. https://doi.org/10.1021/pr700804c

Park, J., & LaBaer, J. (2006). Recombinational cloning. *Current Protocols in Molecular Biology*, *74*(1). https://doi.org/10.1002/0471142727.mb0320s74

Ramachandran, N., Hainsworth, E., Bhullar, B., Eisenstein, S., Rosen, B., Lau, A. Y., Walter, J. C., & LaBaer, J. (2004). Self-assembling protein microarrays. *Science*, *305*(5680), 86–90. https://doi.org/10.1126/science.1097639

Ramachandran, N., Raphael, J. V., Hainsworth, E., Demirkan, G., Fuentes, M. G., Rolfs, A., Hu, Y., & LaBaer, J. (2008). Next-generation high-density self-assembling functional protein arrays. *Nature Methods*, *5*(6), 535–538. https://doi.org/10.1038/nmeth.1210

Exercises 14.1

1. In NAPPA, the IVTT is added?
    a. On each spot
    b. With the cDNA template
    c. As a solution on the slide
    d. On a membrane placed on the slide

2. What is the purpose of PicoGreen dye in a NAPPA microarray experiment?
    a. To check the protein produced on the array
    b. To check the DNA printed on the array
    c. To check whether functional protein has been produced
    d. To check whether DNA translation is occurring

3. How is the DNA attached to an aminosilane-coated slide?
   a. By using biotin-avidin reaction
   b. By bond formation of the amino group and phosphate group of DNA, under IR
   c. By baking at 85°C
   d. No other treatment needed after spotting

4. NAPPA technique is used for which of the following applications?
   a. Biomarker discovery
   b. Protein-protein interactions
   c. Autoantibody screening
   d. All of the above

5. NAPPA acquires its high-throughput capabilities due to which of the following characteristics?
   a. Utilizing expression ready clones from repositories
   b. Utilizing robotics for cloning and array printing
   c. Utilizing cell-free synthesis system for protein expression
   d. All of the above

6. Why is BSA added to the master mix in NAPPA experiments?
   a. It improves binding efficiency of cDNA to the chip
   b. It improves binding efficiency of translated protein to the chip
   c. It improves binding efficiency of aminosilane to the chip
   d. It improves binding efficiency of BS^3 crosslinker to the chip

7. DNA is immobilized on chip surface by UV radiation in which of the following surface chemistries?
   a. Nitrocellulose
   b. Gold
   c. Silane
   d. Glass

8. Which of the following steps must be incorporated while fabricating NAPPA chips?
   a. Chip layout must be decided after chip printing
   b. Positive and negative control spots must be incorporated in chip layout
   c. Array scheme must be fed into the arrayer before chip printing
   d. Only a and b
   e. Only b and c

9. To print quality arrays and avoid day-to-day variations, which of the following precautions must be followed?
   a. Usage of liquid handling systems
   b. Checking fidelity of protocols
   c. Quality control check of spots
   d. All of the above

10. For NAPPA chip printing, it is ideal that a user must have which of the following?
    a. Access to a repository of expression-ready clones
    b. Access to robotic arrayer/printer
    c. Purified DNA templates
    d. Only a and b
    e. a, b, and c

11. You want to generate nucleic acid programmable protein arrays (NAPPA). You have a DNA template, anti-GST antibody attached to the slide surface. Your task is to identify when the protein synthesis happens and mark it on the figure given below. Individual figures of newly synthesized protein and GST tag are given. Where do these small figures fit in the workflow of NAPPA?

Workflow:

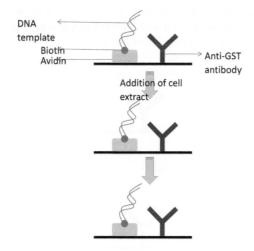

Protein synthesis:

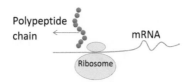

Newly synthesized protein:

GST tag:

Answers

1. c
2. b
3. c
4. d
5. d
6. a
7. c
8. e
9. d
10. e
11. The transcription starts as soon as the cell extract is added to the slide surface. Once the polypeptide/protein is expressed along with the GST tag, this protein tag complex attaches to the anti-GST antibody as shown in the figure below.

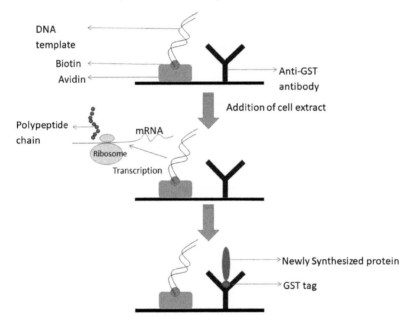

15

Label-Free Proteomics

Preamble

Over the last two decades, detection techniques in proteomics have experienced phenomenal achievements with the introduction of different ultra-sensitive detection techniques, which can selectively detect target analytes from even complex biological samples. Additionally, quite a few detection techniques are proficient for multiplexed detection, which is very useful for high-throughput proteomics applications, particularly in protein/antibody microarrays. There are two major detection approaches used in protein microarrays; label-based and label-free. In label-free detection approaches, inherent properties of the query molecules like mass and dielectric property, are measured, which eliminates any requirement or interference due to the presence of any tagging molecules. Recently, label-free detection methods are gaining recognition due to their easy operation procedure, real-time detection, exclusion of the need of secondary reactants and prolonged labeling practice. This chapter will provide an introduction and working principle of different label-free detection techniques which are commonly applied in proteomics.

15.1 Label-Based Detection Techniques

With rapid advancements in gel-free proteomics techniques, particularly protein microarrays, the need for improved detection systems has been imperative. Label-based detection systems have taken rapid strides to satisfy this demand with significant improvements in sensitivity, multiplexing capability and reproducibility. In case of the chemiluminescence technique, the antigen of interest binds to the corresponding antibodies coated on the array surface (Figure 15.1). The array is then probed by an enzyme-linked secondary antibody that is capable of recognizing a different epitope on the same antigen. The excess unbound antibody is washed off and the chemiluminescent substrate is then added which reacts with the enzyme and emits light. This is detected through a CCD camera and a plot is obtained.

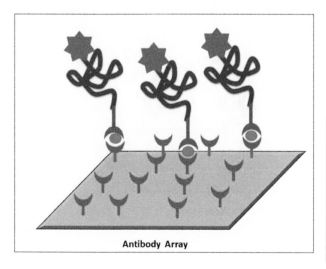
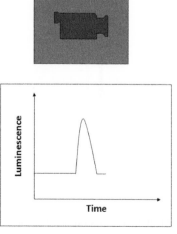

FIGURE 15.1 Schematic for chemiluminescence.

DOI: 10.1201/9781003098645-19

The fluorescence-based detection technique uses an array surface that is functionalized with probe antibody molecules specific to the target antigen of interest (Figure 15.2). The target antigens are bound to their primary antibodies on the array surface. Detection is carried out using fluorescent-labeled secondary antibodies. The excess unbound secondary antibody is washed off, and the fluorescence is measured by exciting the array with the light of a suitable wavelength. The resulting emission is measured using a microarray scanner and can be used to quantify the corresponding antigen-antibody interaction. Sensitivities of less than 1 ng are achievable by these fluorescent dyes.

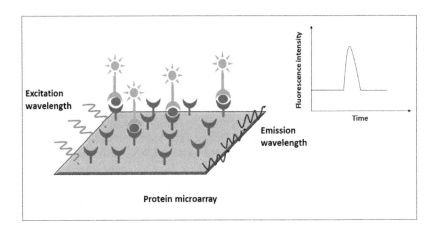

FIGURE 15.2   Schematic for fluorescence-based detection.

Antigen-antibody binding interaction can easily be detected by means of chromogenic reactions (Figure 15.3). An enzyme that can give a colored reaction upon adding a suitable substrate molecule is linked to the secondary antibody. This acts as a probe by binding to a different epitope on the antigen from that of the primary antibody bound to the array surface. The binding of the substrate molecule results in the colored product being formed, which is easily detected and quantified through an array scanner. This detection technique has achieved sensitivity down to femtomolar levels.

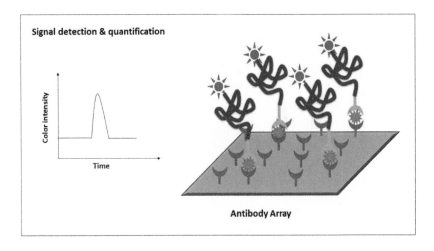

FIGURE 15.3   Schematic for chromogenic detection.

In the case of radioactive labeling, the array surface is coated with the protein mixture containing the target protein of interest. A suitable radio-labeled query protein that can specifically interact with the protein of interest is used to probe the array surface. Once the binding has occurred, excess unbound query protein is washed off the surface. The washed array surface is then developed in an autoradiography solution. Beta emissions from the radioactive carbon atoms of the query protein strike the photographic film on which the final image is then developed (Figure 15.4).

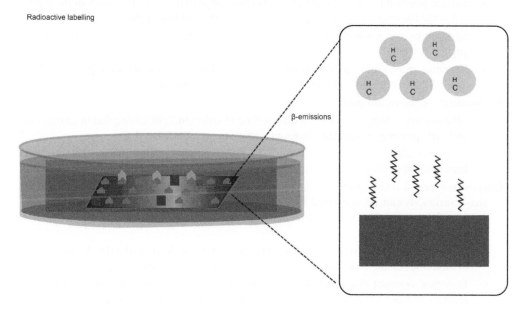

FIGURE 15.4 Schematic for radioactive labeling.

15.2 Label-Free Detection Techniques Commonly Used in Proteomics

The development of reliable, sensitive, and high-throughput label-free detection techniques has become imperative for proteomic studies due to drawbacks associated with label-based technologies. Label-free detection methods (Figure 15.5), which monitor the inherent properties of the query molecule, promise to simplify bioassays. These techniques are briefed as under.

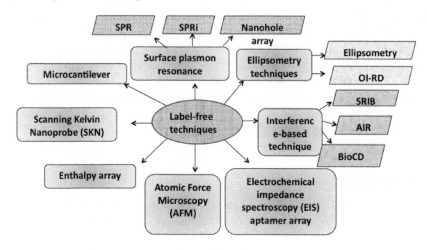

FIGURE 15.5 An overview of label-free techniques.

Overview of label-free techniques: Label-free detection methods keep track of the molecules' inborn characteristics, including their mass, optical properties, and dielectric properties. In contrast to label-based detection methods, these methods do not tag the query molecules, preventing changes in structure and function. Despite not requiring time-consuming procedures, they have drawbacks, such as sensitivity and specificity problems.

Surface plasmon resonance-based techniques:

   i. Surface plasmon resonance (SPR): Detects any change in the refractive index of the material at the interface between the metal surface and the ambient medium.

   ii. Surface plasmon resonance imaging (SPRi): Image reflected by polarized light at a fixed angle is detected.

   iii. Nanohole array: Light transmission of specific wavelength enhanced coupling surface plasmons (SPs) on both sides of metal surface with periodic nanoholes.

Ellipsometry-based techniques:

   i. Ellipsometry: Change in the polarization state of reflected light arising due to changes in dielectric property or refractive index of surface material measured.

   ii. Oblique incidence reflectivity difference (OI-RD): Variation of ellipsometry that monitors harmonics of modulated photocurrents under nulling conditions.

Interference-based techniques: The foundation of interferometry is the idea that wave front phase variations can be converted into interference fringes, which are easily measurable intensity fluctuations. The various detection strategies that make use of this principle include the following:

   i. Spectral reflectance imaging biosensor (SRIB): Changes in the optical index due to capture of molecules on the array surface detected using optical wave interference.

   ii. Biological compact disc (BioCD): Local interferometry, i.e., transformation of phase differences of wave fronts into observable interference fringes, used for detection of protein capture.

   iii. Arrayed imaging reflectometry (AIR): Destructive interference of polarized light reflected from silicon substrate captured and used for detection.

Electrochemical impedance spectroscopy (EIS) – aptamer array: Short single-stranded oligonucleotides called aptamers have a broad range of target biomolecules to which they can bind. EIS and aptamer arrays together can provide a highly sensitive label-free detection method.

Atomic force microscopy (AFM): Vertical or horizontal deflections of cantilever measured by high-resolution scanning probe microscope, thereby providing significant information about surface features.

Enthalpy array: Thermodynamics and kinetics of molecular interactions measured in small sample volumes without any need for immobilization or labeling of reactants.

Scanning Kelvin nanoprobe (SKN): A non-contact technique that does not require specialized vacuum or fluid cell; SKN detects regional variations in surface potential across the substrate of interest caused due to molecular interactions.

Microcantilever: Thin, silicon-based, gold-coated surfaces suspended on a sturdy framework are called microcantilever. Bending of the cantilever due to surface adsorption is detected either electrically by metal oxide semiconductor field-effect transistors or optically by changes in the angle of reflection.

15.2.1 SPR and Related Techniques

SPR is a label-free technique that measures variations in the refractive index of the dielectric layer adjoining the sensor surface as a result of the adsorption or desorption of molecules (Figure 15.6). SPR provides real-time measurements of alteration in refractive index in the locality of a surface. The variation in reflection intensity to an incident angle before and after binding of the target molecule is shown as sensorgram.

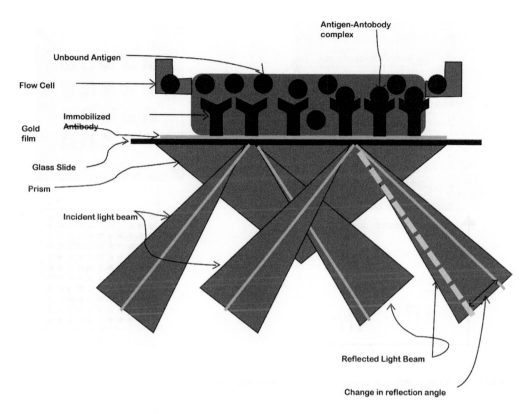

Unbound Antigen

Antigen-Antobody complex

Flow Cell

Gold film

Immobilized Antibody

Glass Slide

Prism

Incident light beam

Reflected Light Beam

Change in reflection angle

FIGURE 15.6 Representation of principle of SPR technique where variations in the refractive index of the medium directly in contact with the sensor surface are measured.

In SPRi, a spatially resolved measuring device is introduced in the SPR set-up (Ladd et al., 2009). SPR and SPRi are suitable for instantaneous label-free analysis of several biomolecular interactions in a quick and HT style. SPR-based biosensors capable of detecting very minute amounts of target analytes with high selectivity, are promising for discovering disease biomarkers. Apart from SPR and SPRi, nanohole arrays are considered as an advantageous label-free approach for bio-sensing, since they require plain optical alignment and simple miniaturization and offer high accuracy, robustness, increased fluorescent signal, multiplexing and collinear optical detection (Ji et al., 2008; Lesuffleur et al., 2008). If prearranged arrays of nanoscale holes are designed in a metal film, unusual optical transmission characteristics at resonant wavelengths are observed. SPs are excited on both sides of the metal surface. It increases the light transmission for a specific wavelength and makes nanohole arrays a prospective surface-based biosensor (Figure 15.7).

15.2.2 Ellipsometry-Based Techniques

Ellipsometry-based label-free detection methods measure the polarization state of the reflected light, which is changed when the dielectric property or refractive index of the sample surface is altered (Figure 15.8). In imaging ellipsometry microscopy, the CCD camera is coupled with an ellipsometer (Jin et al., 2004). If the microfluidic system is coupled with imaging ellipsometry, multiple advantages such as high automation, less sample consumption, fast detection, and HT assays with superior sensitivity can be achieved. Ellipsometry-based techniques are useful for studying the kinetics of biomolecular interactions, hormonal activity, detection of microorganisms, and quantification of competitive adsorption of protein. Another form of ellipsometry, OI-RD, in which the harmonics of modulated photocurrents are measured under appropriate nulling conditions, is used as a label-free detection platform in proteomics (Fei et al., 2008; Zhu et al., 2007). Variation in thickness and/or dielectric response as a result of biomolecular interactions generates a detectable OI-RD signal.

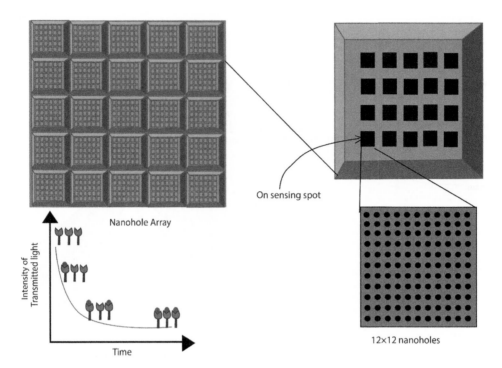

FIGURE 15.7   A representation of nanohole array-based label-free approach for bio-sensing.

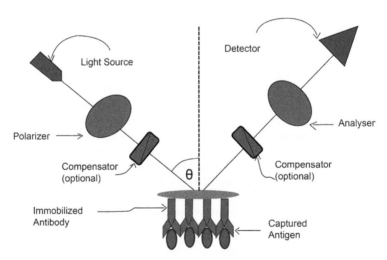

FIGURE 15.8   The basic principle behind ellipsometry. A photodetector is used to monitor the intensity of the reflected light.

15.2.3 Interference-Based Techniques

Interference-based techniques detect optical phase differences as a result of biomolecular mass addition (Figure 15.9). There are quite a few potential interferometric techniques such as SRIB, dual-channel biosensor, SPR interferometry, on-chip interferometric backscatter detection, porous silicon-based optical interferometric biosensor, BioCD and spinning disc interferometry, which is very promising for label-free detection of biomolecules (Ray et al., 2010). Biochemical and functional analysis of proteins is also possible using interference-based label-free detection methods (Table 15.1) (Gao et al., 2006; Ozkumur et al., 2008).

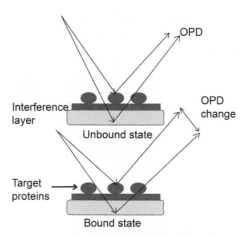

FIGURE 15.9 Showing the interferometric techniques that measure the phase differences of the wave fronts and convert them into observable visible intensity fluctuation known as interference fringes.

A SiO_2-coated Si surface is functionalized with the biomolecule of interest. The magnitude of total reflected light at a particular wavelength depends entirely on the OPD between the top surface and the SiO_2-Si interface. The binding of the target to the immobilized biomolecule further increases the OPD and is seen as a shift in the spectral reflectivity. SRIB, therefore, serves as a useful tool for HT, real-time detection of biomolecular interactions. SRIB is the most promising interference-based label-free detection method, which monitors alterations in the optical index due to the capture of biological material on the sensor surface (Ozkumur et al., 2008) (Figure 15.10). SRIB directly monitors primary molecular binding interactions with high sensitivity. Back-scattering interferometry (BSI) is another promising platform for studying label-free molecular interactions with minimal samples (Bornhop et al., 2007). It can quantify a wide dynamic range of molecular interactions in free solution and very compatible with multiplexing.

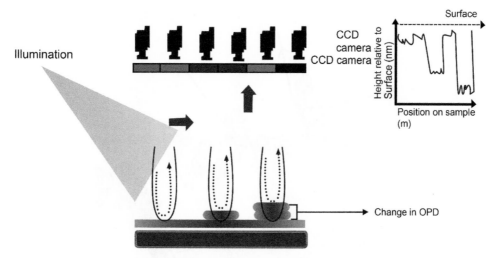

FIGURE 15.10 Spectral reflectance imaging biosensor (SRIB).

15.2.4 Scanning Kelvin Nanoprobe

The Kelvin probe force microscope (KPFM) measures local changes in surface potential across a substrate (generally gold) (Figure 15.11). KPFM is very promising for label-free biomolecular label-free detection and offers several advantages. It is a non-contact technique; therefore, specialized vacuum or fluid cell is not required. High-speed screening is possible with KPFM while maintaining the signal fidelity. Another significant aspect of KPFM technology is its ability to analyze high-density arrays (Sinensky & Belcher, 2007). This label-free detection technique also can reduce noise by decreasing the non-specific binding of biomolecules.

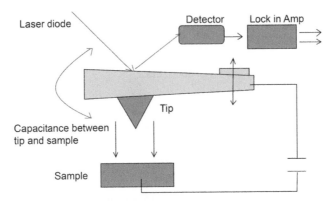

FIGURE 15.11   Working principle of scanning Kelvin nanoprobe. Variations in work function and surface potential occurred due to molecular interactions are measured using SKN.

15.2.5 Microcantilever

Microcantilevers are silicon-based, gold-coated, thin (1 mm) surfaces, horizontally attached to a solid support (Braun et al., 2009). The binding of biomolecules on the cantilevers bends them and the level of bending is measured optically or electrically for label-free detection (Figure 15.12). Analysis of thermodynamics of protein-protein and other biomolecular interactions, detection of cancer markers and antigen-antibody interactions can be performed using microcantilever-based label-free sensors (Table 15.1) (Ray et al., 2010).

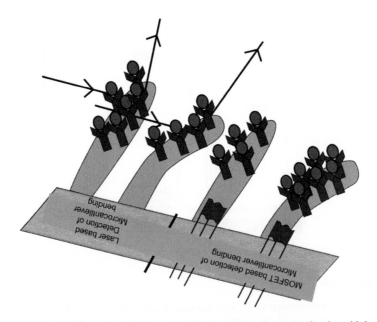

FIGURE 15.12   Working principle of the microcantilever. The interaction of query molecules with immobilized target molecules leads to the bending of microcantilevers and changes the resonant frequency.

TABLE 15.1

Application of Label-Free Techniques in Proteomics

Techniques	Principle	Applications	Merits	Demerits	Sensitivity and Resolution
1. Surface plasmon resonance (SPR) and related techniques					
(i) Surface plasmon resonance (SPR)	Measures variations in the medium's refractive index that are in direct contact with the sensor surface	Drug discovery Studying association or dissociation kinetics Antigen-antibody interactions in protein microarrays Rapid diagnosis of cancer patients	1. Real-time measurements 2. Multiplex analysis 3. Sensitive to conformational change 4. Quantitative and qualitative	Restricted to gold/silver surfaces	10 ng/mL for casein by LSPR High [B]
(ii) Surface plasmon resonance imaging (SPRi)	Captures an image reflected by polarized light at a fixed angle, and simultaneously detects many biomolecular interactions	Protein-glycan interactions Used for DNA-protein interaction Antigen-antibody interactions on microarrays disease marker detection and protein expression profiling	1–4 as above+ 5. Suitable for HT	1. Restricted to gold/silver surfaces 2. Requires sophisticated instrumentation	[nM-zM range] 0.5 pM 64.8 zeptomole [Best achievable sensitivity] Very High [A]
(iii) Nanohole array	Periodic nanoholes couple incident photons into surface plasmons (SPs). SPs of both the side couple through periodic nanoholes to enhance light transmission	Protein-protein interaction Binding kinetics measurement	1,2,4,5 as above+ 6. Simple optical alignment 7. Unlike SPR, bulky prism is not required 8. Use of high numerical aperture is possible 9. Miniaturization is possible	1. Insensitive to conformational changes 2. Restricted to gold/silver surfaces	333 nm/RIU • 9.4×10^{-8} RIU •• 80 nM High [A]
2. Ellipsometry	Measures change in polarization state of the incident light which depends on the dielectric properties and refractive index of thin film	Intrinsic pathway of coagulation Real-time and end-point measurement of biomolecular interactions Clinical diagnosis and narcotics detection Hormone detection and cancer marker test Affinity determination	1,2,4 as above+ 10. Not restricted to gold/silver 11. Cheaper than SPR-based biosensors 12. Simple instrumentation 13. Large field of view for simultaneous monitoring of the entire microarray	1. Less sensitive than SPRi 2. Insensitive to conformational changes	1 ng/mL [51, 56] 10 pg/mm² High [B]
3. Oblique incidence reflectivity difference (OI-RD)	Based on polarization modulated nulling ellipsometry	Real-time and end point analysis of antigen-antibody interaction DNA-DNA hybridization and protein-small-ligand binding reactions	All the merits of ellipsometry +5+ 14. Higher sensitivity than imaging ellipsometry	Insensitive to conformational changes	10 pm thickness change. Sensitivity comparable to SPRi Very High [A]
4. Interference-based techniques					
(i) Spectral reflectance imaging biosensor (SRIB)	Detection of optical phase difference due to biomolecular mass accumulation	Dynamic measurements of protein-protein interactions	5+ 15. Cost effective 16. Fast determination of binding kinetics	1. Suitable for only smooth layered substrates 2. Non-specific binding	19 ng/mL High [B]
(ii) Dual-channel biological compact disc (BioCD)	Simultaneous interferometry and fluorescence detection	Detection of mass and fluorescence signals from protein	5+17 Extremely fast 18. Specific and non-specific bindings can be differentiated	Expensive and complex	30–70 pg/mL High [A]

(Continued)

TABLE 15.1 (*Continued*)

Application of Label-Free Techniques in Proteomics

Techniques	Principle	Applications	Merits	Demerits	Sensitivity and Resolution
(iii) Arrayed imaging reflectometry (AIR)	Measures small-localized changes in optical thickness of a thin film	Detection of human proteins in cellular lysate and serum Biomolecular binding Protein spot homogeneity evaluation	6+15+17	Sensitivity	250 pg/mL, High [A]
5. Scanning Kelvin nanoprobe (SKN)	Measures alteration in work function and surface potential due to molecular interactions	Antigen–antibody interactions DNA structure analysis Isoelectric point determination	17+19. Non-contact	Unsuitable for very complex samples	<50 nanomolar High [B]
6. Atomic force microscope (AFM)	High resolution scanning probe microscope detects vertical and horizontal deflection of cantilever	Pathogen detection Protein interaction	20. Detection under physiologically relevant conditions 21. High specificity	1. Imaging in aqueous solutions is very difficult 2. Image artifacts	Picoliter volume High [B]
7. Enthalpy array	Arrays of nanocalorimeters, measures heat generation of the reaction	Enzyme kinetics (Km, Kcat) and inhibitor constants (Ki) can be determined Biomolecular interactions enzymatic turnover and mitochondrial respiration determination	10+ 22. Immobilization of biomolecules not required 23. Very rapid, small sample volumes required 24. Can be used for complex samples (i.e., serum)	1. False positives when 2 reacting solutions have different pH or ionic strength 2. Complex instrumentation 3. Real-time analysis not possible 4. Not sensitive to conformational change	μM-nM range, Moderate [C]
8. Microcantilevers	The binding of query molecules to the immobilized target molecules cause bending of microcantilever and change the resonant frequency	Investigating thermodynamics of bio-molecular interactions Detecting conformational changes Determining mass of single virus or bacterium and measurement of cell growth on cantilever surface	1,4,10,15	False positives with complex sample (i.e., serum)	0.2 ng/mL High [B]

Source: Modified from (Ray et al., 2010)

Notes: Sensitivity scale

[A] Very high: atto-femtogram/mL [10^{-18}–10^{-15} g/mL]

[B] High: pico-nanogram/mL [10^{-12}–10^{-9} g/mL]

[C] Moderate: microgram/mL [10^{-6} g/mL]

HT applications demonstrated/proof of concept

• Amount of protein used in the microchannel during incubation 17.5 μg.

•• Corresponds to 200 μL of 290 nM GST (~34.9×10^{12} molecule).

15.3 Application of Label-Free Techniques in Proteomics

Several label-free approaches, like SPR, SPRi, interference-based techniques, microcantilever, etc., are considered as impending platforms for studying biomolecular interactions and detection of disease biomarkers (Table 15.1).

15.4 Challenges

Regardless of rapid advancements in the field of label-free proteomics achieved with the introduction of new and versatile technologies, label-free detection approaches have several limitations. Both label-free and label-based detection methods have advantages and limitations (Chandra & Srivastava, 2010; Ray et al., 2010). Although label-free detection techniques are very promising and potential candidates for real-time measurements of low-abundance analytes and protein-protein interactions, issues regarding sensitivity and specificity remain to be explored further. Additionally, costly fabrication techniques, morphological anomalies of sample spots, and insufficient knowledge regarding the exact working principles of the label-free biosensors, often restrict their use for practical clinical applications. Label-free measurements have capabilities of detection of low-abundance analytes and protein-protein interactions, but further improvement of specificity and sensitivity is required when complex body fluids have to be analyzed rather than simple buffer solutions commonly used in proof-of-principle experiments. To make label-free sensors popular in routine clinical applications, cost-effective fabrication techniques are required to be developed, and the mechanism of working principle of label-free detection approaches needs to be explored further and transitioned into clinics.

15.5 Conclusions

Label-free detection techniques are attractive for the large-scale, real-time analysis of protein-protein and other biomolecular interactions and measurement of concentrations of multiple target molecules in HT manner. Such extremely sensitive, fast, label-free detection approaches are useful in various applied fields including pharmaceutical analysis, screening of potential drug molecules, cellular detection, characterization of biomolecules, detection of disease markers and environmental monitoring. The coupling of microarrays and label-free techniques is emerging rapidly and has been found to be highly effective in the detection of extremely low-abundance analytes in buffer solutions. Nonetheless, sensitivity and specificity frequently become the prime limitation for label-free detection methods when very complex biological samples are concerned. Hitherto, label-free detection approaches found to be efficient in the analysis of antigen-antibody interactions. Still they will be useful in actual bed-side applications in clinics if they can detect multiple protein-protein interactions simultaneously with similar efficacy. Considering the present scenario, it can be concluded that label-free proteomics is still at a premature stage of development and has shown promises mainly for targeted detection of known protein parkers; however, it has not contributed effectively in the discovery of new markers which can be directly translated in clinics. It is anticipated that with efforts from different research groups worldwide, the field of label-free proteomics will become more robust, sensitive, fast, cost-effective, and overcome existing limitations.

REFERENCES

Bornhop, D. J., Latham, J. C., Kussrow, A., Markov, D. A., Jones, R. D., & Sørensen, H. S. (2007). Free-solution, label-free molecular interactions studied by back-scattering interferometry. *Science*, *317*(5845), 1732–1736. https://doi.org/10.1126/science.1146559

Braun, T., Ghatkesar, M. K., Backmann, N., Grange, W., Boulanger, P., Letellier, L., Lang, H.-P., Bietsch, A., Gerber, C., & Hegner, M. (2009). Quantitative time-resolved measurement of membrane protein–ligand interactions using microcantilever array sensors. *Nature Nanotechnology, 4*(3), 179–185. https://doi. org/10.1038/nnano.2008.398

Chandra, H., & Srivastava, S. (2010). Cell-free synthesis-based protein microarrays and their applications. *PROTEOMICS, 10*(4), 717–730. https://doi.org/10.1002/pmic.200900462

Fei, Y. Y., Landry, J. P., Sun, Y. S., Zhu, X. D., Luo, J. T., Wang, X. B., & Lam, K. S. (2008). A novel high-throughput scanning microscope for label-free detection of protein and small-molecule chemical microarrays. *Review of Scientific Instruments, 79*(1), 013708. https://doi.org/10.1063/1.2830286

Gao, T., Lu, J., & Rothberg, L. J. (2006). Biomolecular sensing using near-null single wavelength arrayed imaging reflectometry. *Analytical Chemistry, 78*(18), 6622–6627. https://doi.org/10.1021/ac0609226

Ji, J., O'Connell, J. G., Carter, D. J. D., & Larson, D. N. (2008). High-throughput nanohole array based system to monitor multiple binding events in real time. *Analytical Chemistry, 80*(7), 2491–2498. https://doi. org/10.1021/ac7023206

Jin, G., Zhao, Z.-Y., Wang, Z.-H., Meng, Y.-H., Ying, P.-Q., Chen, S., Chen, Y.-Y., Qi, C., & Xia, L.-H. (2004). The development of biosensor with imaging ellipsometry. *The 26th Annual International Conference of the IEEE Engineering in Medicine and Biology Society, 3*, 1975–1978. https://doi.org/10.1109/ IEMBS.2004.1403583

Ladd, J., Taylor, A. D., Piliarik, M., Homola, J., & Jiang, S. (2009). Label-free detection of cancer biomarker candidates using surface plasmon resonance imaging. *Analytical and Bioanalytical Chemistry, 393*(4), 1157–1163. https://doi.org/10.1007/s00216-008-2448-3

Lesuffleur, A., Im, H., Lindquist, N. C., Lim, K. S., & Oh, S.-H. (2008). Laser-illuminated nanohole arrays for multiplex plasmonic microarray sensing. *Optics Express, 16*(1), 219. https://doi.org/10.1364/ OE.16.000219

Ozkumur, E., Needham, J. W., Bergstein, D. A., Gonzalez, R., Cabodi, M., Gershoni, J. M., Goldberg, B. B., & Ünlü, M. S. (2008). Label-free and dynamic detection of biomolecular interactions for high-throughput microarray applications. *Proceedings of the National Academy of Sciences of the United States of America, 105*(23), 7988–7992. https://doi.org/10.1073/pnas.0711421105

Ray, S., Mehta, G., & Srivastava, S. (2010). Label-free detection techniques for protein microarrays: Prospects, merits and challenges. *Proteomics, 10*(4), 731–748. https://doi.org/10.1002/pmic.200900458

Sinensky, A. K., & Belcher, A. M. (2007). Label-free and high-resolution protein/DNA nanoarray analysis using Kelvin probe force microscopy. *Nature Nanotechnology, 2*(10), 653–659. https://doi.org/10.1038/ nnano.2007.293

Zhu, X., Landry, J. P., Sun, Y.-S., Gregg, J. P., Lam, K. S., & Guo, X. (2007). Oblique-incidence reflectivity difference microscope for label-free high-throughput detection of biochemical reactions in a microarray format. *Applied Optics, 46*(10), 1890–1895. https://doi.org/10.1364/ao.46.001890

Exercises 15.1

1. Which of the following approaches is not a label-free method?

   a. SPR

   b. Ellipsometry

   c. Scanning Kelvin nanoprobe

   d. Epitope tagging

2. Which of the following approaches is a label-free method?

   a. Epitope tagging

   b. Ellipsometry

   c. Fluorescent labeling

   d. Biotin-based one-color labeling

3. Which of the following properties are advantages for a label-free detection method?
   a. Elimination of the necessity of secondary reactants
   b. Real-time detection
   c. High sensitivity
   d. All of the above

4. Which of the following label-free techniques is most commonly used for studying biomolecular interactions?
   a. SPR and SPRi
   b. Ellipsometry
   c. Scanning Kelvin nanoprobe
   d. Microcantilevers

5. Which of the following properties is measured in interference-based techniques?
   a. Mass differences
   b. Change in refractive index
   c. Change in conductivity
   d. Optical phase differences

6. OI-RD stands for?
   a. Optical-incidence response difference
   b. Optical-incidence reflectivity diffraction
   c. Oblique-incidence response difference
   d. Oblique-incidence reflectivity difference

7. Diffraction-based biosensor differs from SPR as it _____.
   a. Provides real-time measurements
   b. Depends on the diffractive beam pattern
   c. Can perform label-free biomolecular interaction analysis
   d. None of the above

8. Which one of the following is a demerit of nanohole arrays?
   a. It cannot perform multiplex analysis
   b. It cannot perform real-time measurements
   c. It is sensitive to conformational changes
   d. None of the above

9. What is the output from an SPR machine called?
   a. Sonogram
   b. Sensorgram
   c. Thermogram
   d. Rheogram

10. Which of the following is *not* true about label-based techniques?
    a. Require labeling of the query molecule with a marker tag
    b. It is a widely used approach
    c. Tags are easy to attach and do not interfere with the molecular function
    d. Typically endpoint measurement assays

11. In label-free detection methods, the inherent properties of the query molecule such as mass, dielectric property, etc., are measured. This type of measurement ensures the elimination of inference, which may arise due to the presence of tagging molecules. There are various kinds of label-free detection techniques, which are employed in the field of proteomics. Based on the figures given below, name the label-free detection methods and briefly describe the working principles.

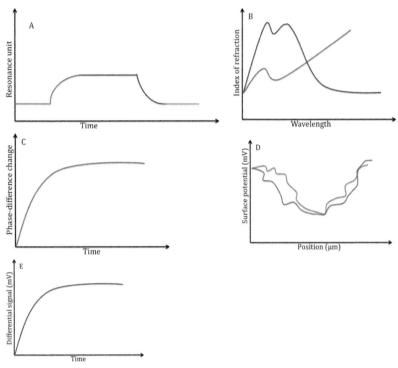

Answers

1. d
2. b
3. d
4. a
5. d
6. d
7. b
8. c
9. b
10. c
11.

    A. **Surface Plasmon Resonance (SPR):** The working principle of SPR is based on the transformation of the energy from the monochromatic incident light beam into electromagnetic energy leading to the production of evanescent waves. This transformation is attributed to the light beam hitting the metal-dielectric interface at a particular SPR angle. So when biomolecules such as proteins are immobilized on gold-surface and interact with their complementary molecule, the angle of reflection of light is altered. This change in the angle of reflection of light is caused by binding of the complementary molecule to the immobilized protein.

B. **Ellipsometry-based techniques:** These techniques measure the polarization state of the reflected light that is changed upon the alteration in dielectric property or index of refraction of the sample surface.

C. **Interference-based techniques:** The basic working principle of interference-based techniques depends on the transformation of phase differences of wave fronts into observable intensity fluctuations known as interference fringes.

D. **Scanning Kelvin nanoprobe:** In this method, the Kelvin probe force microscope measures of local changes in surface potential across a substrate such as gold.

E. **Microcantilever:** The basic principle of microcantilever is based on the binding of the biomolecules to the cantilevers. This binding leads to the bending of the cantilevers and the degrees of bending are measured optically or electrically. Such measurement enables the label-free detection of biomolecules.

16

Surface Plasmon Resonance

Preamble

Surface plasmon resonance (SPR) is a very promising and widely used label-free technique for studying molecular interactions. This label-free technique measures alterations in the refractive index of the dielectric layer adjacent to the metal film owing to the adsorption or desorption of molecules on the surface in real-time mode. SPR is suitable for label-free analysis of several biomolecular interactions in a fast and high-throughput manner. SPR-based biosensors are capable of detecting extremely minute amounts of target analytes with high selectivity. Hence it is very promising for the discovery of disease biomarkers. An introduction to SPR, its working principle, versatile applications, and advantages and disadvantages will be discussed in this chapter.

16.1 Basic Working Principle of Surface Plasmon Resonance

Surface plasmon resonance (SPR) happens when energy from a monochromatic incident light beam hits the metal-dielectric interface at a particular SPR angle (Englebienne et al., 2003) and becomes transformed into electromagnetic energy resulting in the production of evanescent waves (Figure 16.1). Gold surface is generally used in SPR for immobilization of test proteins. The unlabeled query molecules are introduced in solution form and alterations in the angle of reflection of light due to the binding of the probes to the immobilized protein provide real-time information about biomolecular interactions.

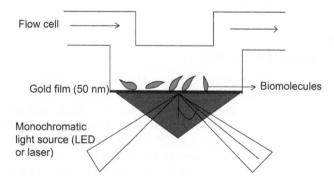

FIGURE 16.1 The basic principle of surface plasmon resonance (SPR). The measurement of alterations in the refractive index of medium directly in contact with the sensor surface.

SPR is a highly sensitive spectroscopic tool that is increasingly being used for label-free detection studies. Test proteins such as antibodies are immobilized onto the gold-coated glass array surface. Incident light striking the surface is constantly reflected at a particular angle in this state. Unlabeled free antigens or other query proteins enter via the flow cell and move toward the immobilized antibodies or other test proteins. There is no change in reflected light upon entering into the system. The binding of antigen to antibody immediately changes in the angle of reflection of light due to changes in the refractive index of the medium. The angle at which the minimum intensity of the reflected light is

DOI: 10.1201/9781003098645-20

achieved is called the "SPR angle". SPR angle is directly related to the number of biomolecules bound to the sensor surface (Figure 16.2). Different factors such as the nature of the metal layer, angle of SPR, the refractive index at the metal-dielectric interface, the wavelength of the incident light, etc., regulate the magnitude of surface plasmon resonance.

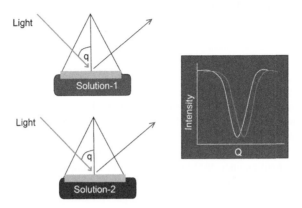

FIGURE 16.2   Angle of reflection in surface plasmon resonance.

In SPR measurement, the sensorgram indicates the changes in reflection intensity to an incident angle before and after binding of the target molecule. Overall SPR-based biosensors contain the incident light source, sensor gold surface, and the detector to capture the reflected light. The interaction between the molecules is measured by plotting the sensorgram (as shown in panels from top to bottom) (Figure 16.3). The ligand molecules are immobilized on an activated gold sensor chip, while the query molecule in the buffer flows through the flow cell. Interaction along with binding kinetics is monitored in a real-time manner.

In SPR, the changes can be continuously monitored to characterize biomolecular interactions in real-time. The SPR angle, i.e., the angle at which minimum intensity of reflected light is obtained is indicative of the amount of biomolecule binding to the surface. The graph represents the change in reflection intensity before and after the antigen-binding (Figure 16.4).

16.2 Different Applications of SPR

In the last 10 years, several research groups have used SPR and related label-free techniques for real-time analysis of protein-protein and other biomolecular interactions, measurement of low abundance serum biomarkers, and screening of inhibitors of tumor targets and potential drug molecules (Table 16.1) (Ray et al., 2010; Reddy et al., 2012). SPR has also been applied extensively for many biomedical, food and environmental applications (Shankaran et al., 2007). Ultra-sensitive detection is required for the measurement of very low-abundance biomarkers. In a study, Choi et al. utilized an SPR-based biosensor for the detection of prostate-specific antigen (PSA) (Choi et al., 2008). In this study, the authors have employed gold (Au)-nanoparticle-antibody complex for signal enhancement of SPR, thereby effectively increased the sensitivity of the detection approach. SPR-based immune-sensor has been designed, where a gold surface coated with PSA monoclonal antibodies (mAbs) and gold-nanoparticle-conjugated antibody complex was used to capture PSA antigen (Figure 16.5). With this SPR-based biosensor, the authors could detect PSA with a detection limit of 300 fM. On a gold sensor device functionalized with recombinant protein G via thiol groups, PSA mAbs were coated. The sensorgram shows the target antigen's interaction with immobilized antibodies. To track these interactions in intricate clinical samples, the SPR angle change proved insufficient. The signal was enhanced by a sandwich immunoassay technique using gold nanoparticles coated with a PSA polyclonal antibody complex. The level of sensitivity was greatly increased by using the gold-nanoparticle-antibody complex as a signal amplifier.

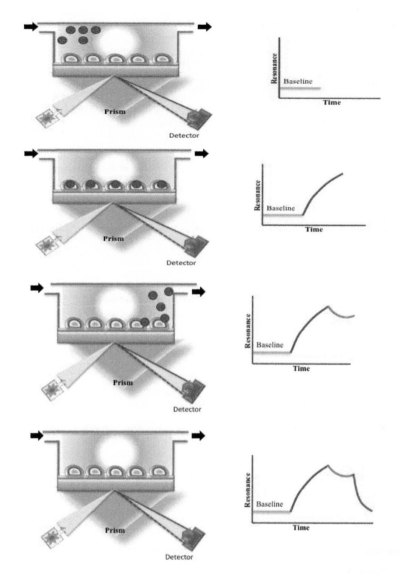

FIGURE 16.3 Different steps involved in biomolecular interaction analysis using SPR sensor.

16.3 Advantages and Disadvantages

SPR has its advantages and limitations like other technologies.

Advantages

- Label-free detection eliminates the necessity of any secondary reactants and lengthy labeling process.
- Real-time measurements of biomolecular interactions.
- The multiplex analysis is possible; therefore, compatible with high-throughput assays.
- Sensitive to conformational changes; direct measurements and study of binding kinetics are possible.
- Both quantitative and qualitative measurements are possible.

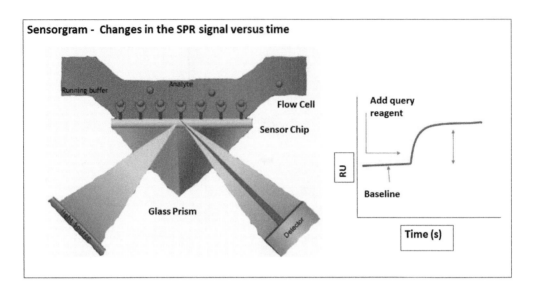

FIGURE 16.4   Representation of sensogram obtained in surface plasmon resonance.

Disadvantages

- Requires sophisticated instrumentation.
- Restricted to choice of metal (gold/silver surfaces).
- Completely dependent on mass changes.
- Temperature-sensitive; change in temperature affects refractive index.
- Non-specific interactions affect the signal.
- Works efficiently only with a homogeneous surface.
- Sensitivity and specificity often become major concerns while handling very complex biological samples.

16.4 Conclusions

The SPR-based label-free sensors are very promising for the detection of disease biomarkers and real-time screening of biomolecules. However, SPR and related technologies have limited usage for large-scale applications in industries and clinics due to multiple technical limitations discussed above. Current advancements in the field of SPR have introduced quite a few new materials and methods for the improvement of the sensitivity of the instruments (Halpern et al., 2011; Reddy et al., 2012). Application of different polarization methods for the incident light (such as p-polarized, s-polarized, TM waves, and TE waves) can effectively enhance the coupling of incident light to plasmons. Besides, there are constant efforts to make reproducible and well-established surface chemistry for the generation of a selective sensing interface, which can reduce the binding of non-specific moieties that can alter the SPR signal. Multi dimensional applications of SPR-based techniques have been achieved through successful coupling with different technological approaches, including SPR-MS, electrochemical SPR, and surface plasmon fluorescence spectroscopy which effectively circumvents some of the basic limitations associated with SPR. These efforts can expand the applications of SPR-based biosensors in clinical diagnosis.

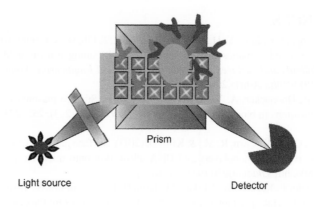

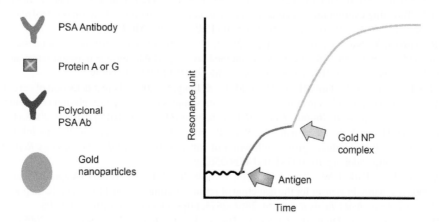

FIGURE 16.5 Sandwich immunoassay SPR platform for detection of cancer biomarker; prostate-specific antigen (PSA) (Choi et al., 2008).

TABLE 16.1

Different Applications of SPR-Based Biosensors (Selected Examples)

Application	Studies (References)
(a) Study of biomolecular interactions	1. Antigen-antibody interactions (Hiep et al., 2007)
	2. Protein-glycan interactions (Yuk et al., 2006)
	3. DNA-protein interaction (Zhu et al., 2000)
	4. Studying association or dissociation kinetics (Wassaf et al., 2006)
(b) Detection of cancer biomarkers	1. Pancreatic cancer: CD166 (Vaisocherová et al., 2009)
	2. Hepatocellular tumors: α-fetoprotein (Teramura & Iwata, 2007)
	3. Ovarian cancer: cancer antigen 125 (Suwansa-ard et al., 2009)
	4. Colon cancer: CEA (Ladd et al., 2009)
(c) Detection of biomarkers for other human diseases	1. Type 2 diabetes: retinol-binding protein 4 (Lee et al., 2008)
	2. Heart diseases: B-type natriuretic peptide (Kurita et al., 2006), myoglobin and cardiac troponin I (Masson et al., 2007)
(d) Screening of inhibitors of tumor targets	1. uPAR-uPA interaction (Khanna et al., 2011)
	2. Aurora B (Lang et al., 2010)
	3. Polo-like kinase 1, Akt, and C-Src (Li et al., 2009; Ma et al., 2011)

REFERENCES

Choi, J.-W., Kang, D.-Y., Jang, Y.-H., Kim, H.-H., Min, J., & Oh, B.-K. (2008). Ultra-sensitive surface plasmon resonance based immunosensor for prostate-specific antigen using gold nanoparticle–antibody complex. *Colloids and Surfaces A: Physicochemical and Engineering Aspects*, *313–314*, 655–659. https://doi.org/10.1016/j.colsurfa.2007.05.057

Englebienne, P., Hoonacker, A. V., & Verhas, M. (2003). Surface plasmon resonance: Principles, methods and applications in biomedical sciences. *Spectroscopy*, *17*(2–3), 255–273. https://doi.org/10.1155/2003/372913

Halpern, A. R., Chen, Y., Corn, R. M., & Kim, D. (2011). Surface plasmon resonance phase imaging measurements of patterned monolayers and DNA adsorption onto microarrays. *Analytical Chemistry*, *83*(7), 2801–2806. https://doi.org/10.1021/ac200157p

Hiep, H. M., Endo, T., Kerman, K., Chikae, M., Kim, D.-K., Yamamura, S., Takamura, Y., & Tamiya, E. (2007). A localized surface plasmon resonance based immunosensor for the detection of casein in milk. *Science and Technology of Advanced Materials*, *8*(4), 331–338. https://doi.org/10.1016/j.stam.2006.12.010

Khanna, M., Chelladurai, B., Gavini, A., Li, L., Shao, M., Courtney, D., Turchi, J. J., Matei, D., & Meroueh, S. (2011). Targeting ovarian tumor cell adhesion mediated by tissue transglutaminase. *Molecular Cancer Therapeutics*, *10*(4), 626–636. https://doi.org/10.1158/1535-7163.MCT-10-0912

Kurita, R., Yokota, Y., Sato, Y., Mizutani, F., & Niwa, O. (2006). On-chip enzyme immunoassay of a cardiac marker using a microfluidic device combined with a portable surface plasmon resonance system. *Analytical Chemistry*, *78*(15), 5525–5531. https://doi.org/10.1021/ac060480y

Ladd, J., Lu, H., Taylor, A. D., Goodell, V., Disis, M. L., & Jiang, S. (2009). Direct detection of carcinoembryonic antigen autoantibodies in clinical human serum samples using a surface plasmon resonance sensor. *Colloids and Surfaces B: Biointerfaces*, *70*(1), 1–6. https://doi.org/10.1016/j.colsurfb.2008.11.032

Lang, Q., Zhang, H., Li, J., Xie, F., Zhang, Y., Wan, B., & Yu, L. (2010). 3-Hydroxyflavone inhibits endogenous Aurora B and induces growth inhibition of cancer cell line. *Molecular Biology Reports*, *37*(3), 1577–1583. https://doi.org/10.1007/s11033-009-9562-y

Lee, S. J., Youn, B.-S., Park, J. W., Niazi, J. H., Kim, Y. S., & Gu, M. B. (2008). SsDNA aptamer-based surface plasmon resonance biosensor for the detection of retinol binding protein 4 for the early diagnosis of type 2 diabetes. *Analytical Chemistry*, *80*(8), 2867–2873. https://doi.org/10.1021/ac800050a

Li, L., Wang, X., Chen, J., Ding, H., Zhang, Y., Hu, T., Hu, L., Jiang, H., & Shen, X. (2009). The natural product Aristolactam AIIIa as a new ligand targeting the polo-box domain of polo-like kinase 1 potently inhibits cancer cell proliferation. *Acta Pharmacologica Sinica*, *30*(10), 1443–1453. https://doi.org/10.1038/aps.2009.141

Ma, J., Huang, H., Chen, S., Chen, Y., Xin, X., Lin, L., Ding, J., Liu, H., & Meng, L. (2011). PH006, a novel and selective Src kinase inhibitor, suppresses human breast cancer growth and metastasis in vitro and in vivo. *Breast Cancer Research and Treatment*, *130*(1), 85–96. https://doi.org/10.1007/s10549-010-1302-4

Masson, J.-F., Battaglia, T. M., Khairallah, P., Beaudoin, S., & Booksh, K. S. (2007). Quantitative measurement of cardiac markers in undiluted serum. *Analytical Chemistry*, *79*(2), 612–619. https://doi.org/10.1021/ac061089f

Ray, S., Mehta, G., & Srivastava, S. (2010). Label-free detection techniques for protein microarrays: Prospects, merits and challenges. *Proteomics*, *10*(4), 731–748. https://doi.org/10.1002/pmic.200900458

Reddy, P. J., Sadhu, S., Ray, S., & Srivastava, S. (2012). Cancer biomarker detection by surface plasmon resonance biosensors. *Clinics in Laboratory Medicine*, *32*(1), 47–72. https://doi.org/10.1016/j.cll.2011.11.002

Shankaran, D., Gobi, K., & Miura, N. (2007). Recent advancements in surface plasmon resonance immunosensors for detection of small molecules of biomedical, food and environmental interest. *Sensors and Actuators B: Chemical*, *121*(1), 158–177. https://doi.org/10.1016/j.snb.2006.09.014

Suwansa-ard, S., Kanatharana, P., Asawatreratanakul, P., Wongkittisuksa, B., Limsakul, C., & Thavarungkul, P. (2009). Comparison of surface plasmon resonance and capacitive immunosensors for cancer antigen 125 detection in human serum samples. *Biosensors and Bioelectronics*, *24*(12), 3436–3441. https://doi.org/10.1016/j.bios.2009.04.008

Teramura, Y., & Iwata, H. (2007). Label-free immunosensing for α-fetoprotein in human plasma using surface plasmon resonance. *Analytical Biochemistry*, *365*(2), 201–207. https://doi.org/10.1016/j.ab.2007.03.022

Vaisocherová, H., Faca, V. M., Taylor, A. D., Hanash, S., & Jiang, S. (2009). Comparative study of SPR and ELISA methods based on analysis of CD166/ALCAM levels in cancer and control human sera. *Biosensors and Bioelectronics, 24*(7), 2143–2148. https://doi.org/10.1016/j.bios.2008.11.015

Wassaf, D., Kuang, G., Kopacz, K., Wu, Q.-L., Nguyen, Q., Toews, M., Cosic, J., Jacques, J., Wiltshire, S., Lambert, J., Pazmany, C. C., Hogan, S., Ladner, R. C., Nixon, A. E., & Sexton, D. J. (2006). High-throughput affinity ranking of antibodies using surface plasmon resonance microarrays. *Analytical Biochemistry, 351*(2), 241–253. https://doi.org/10.1016/j.ab.2006.01.043

Yuk, J. S., Kim, H.-S., Jung, J.-W., Jung, S.-H., Lee, S.-J., Kim, W. J., Han, J.-A., Kim, Y.-M., & Ha, K.-S. (2006). Analysis of protein interactions on protein arrays by a novel spectral surface plasmon resonance imaging. *Biosensors & Bioelectronics, 21*(8), 1521–1528. https://doi.org/10.1016/j.bios.2005.07.009

Zhu, H., Klemic, J. F., Chang, S., Bertone, P., Casamayor, A., Klemic, K. G., Smith, D., Gerstein, M., Reed, M. A., & Snyder, M. (2000). Analysis of yeast protein kinases using protein chips. *Nature Genetics, 26*(3), 283–289. https://doi.org/10.1038/81576

Exercises 16.1

1. Which of the following properties is measured in surface plasmon resonance?
   a. Mass differences
   b. Change in refractive index
   c. Change in conductivity
   d. Optical phase differences

2. SPR can be coupled with which of the following?
   a. Immunoassays
   b. Mass spectrometry
   c. Fluorescence spectroscopy
   d. All of the above

3. Which of the following property(s) are advantages for an SPR-based biosensor?
   a. Label-free detection without the necessity of any secondary reactants
   b. Multiplex analysis is possible
   c. Real-time measurements of biomolecular interaction
   d. All of the above

4. Which of the following label-free techniques is the most potential approach for studying binding kinetics?
   a. SPR and SPRi
   b. Ellipsometry
   c. Scanning Kelvin nanoprobe
   d. Interference-based techniques

5. Enhancement of the coupling of incident light to plasmons is made possible by which of the following?
   a. Application of different polarization methods for the incident light
   b. Elimination of contaminants
   c. Addition of imaging device
   d. None of the above

6. Sensorgram is a plot of _____.
   a. Response against concentration
   b. Response against time
   c. K_D against concentration
   d. K_D against time

7. What is desired as the best optimized regeneration condition in SPR?
   a. All bound analytes are removed
   b. Minimum effect on ligand activity
   c. Both of the above
   d. None of the above

8. In SPR, which one of the following immobilization methods generally provides higher binding capacity?
   a. Direct immobilization
   b. Capture approach
   c. There is no difference between the approaches
   d. None of the above

9. The order of steps in an SPR assay are as follows:
   a. Sample injection, surface preparation, regeneration
   b. Surface preparation, sample injection, regeneration
   c. Sample injection, regeneration, surface preparation
   d. Surface preparation, regeneration, sample injection

10. In an SPR assay, purity of the ligand molecules does not affect the immobilization procedure.
    a. True
    b. False

11. While performing a label-free detection of antigen-antibody binding experiment using surface plasmon resonance (SPR), you realized that the signal intensity is rather low. To boost the signal one can use an approach and the figures shown below are from such experiment. Have a look at these figures and comment on the curves in the figures.

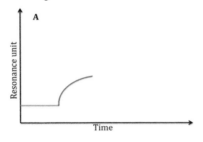 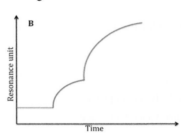

Answers

1. b
2. d
3. d

4. a

5. a

6. b

7. c

8. a

9. b

10. b

11. One can increase the signal intensity by adding a secondary antibody in the assay. The addition of the secondary antibody in the assay would lead to further change in the reflection angle of the light and provides better signal intensity.

   In the figures A and B, the blue and red curves indicate the baseline and the association constant of the primary antibody, respectively. In the figure B, the association constant of the secondary antibody is represented by the green curve and one can observe the increase in the signal intensity for the same antigen-antibody interaction.

17

Surface Plasmon Resonance Imaging

Preamble

Surface plasmon resonance imaging (SPRi) based label-free sensors, which depend on the measurement of an inherent property (refractive index) of query molecules, are gaining popularity in high throughput (HT) proteomics due to their multiplexing capabilities. The basic working principle of SPRi is quite similar to that of SPR, which measures the changes in the refractive index. In SPRi, a spatially resolved imaging device is introduced to SPR set-up for continuously monitoring the changes occurring on the surface to generate real-time kinetic data. SPRi-based biosensors are capable of instantaneous label-free analysis of several biomolecular interactions in a fast and HT manner. In this chapter, an introduction of SPRi, the basic principle behind its operation, different applications, and advantages and disadvantages of SPRi in comparison to SPR will be discussed.

17.1 Basic Working Principle of Surface Plasmon Resonance Imaging

SPRi relies on the measurement of changes in the refractive index of the medium directly in contact with the sensor surface. A gold-coated glass array surface is used for the immobilization of antibodies complementary to the target protein of interest. A broad beam, monochromatic, polarized light originating from a suitable light source is used to illuminate the entire biochip surface with the help of mirrors placed at suitable angles that reflect the light onto the surface. Reflected light from each spot on the array surface is captured through a coupled charge device (CCD) camera and used to generate the SPRi image for simultaneous capturing of reflected light from each spot. The binding of the target antigen to the antibody is detected in real time due to the changes in the intensity of reflected light from every spot on the array surface (Figure 17.1). Multiple biomolecular interactions can be studied simultaneously in a high throughput (HT) manner and changes occurring on the array surface can provide kinetic data about the interactions.

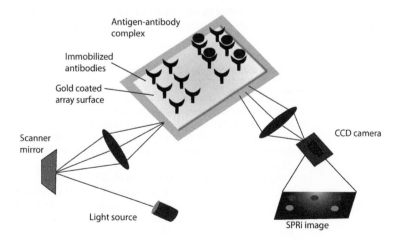

FIGURE 17.1 Basic working principle of surface plasmon resonance imaging (SPRi).

DOI: 10.1201/9781003098645-21

In SPRi, the intensity of the incident light as well as the wavelength are kept constant, and the reflected light is determined at an optimum reflectance angle coming from the metal interface. The complete array can be captured by CCD camera for HT studies. SPRi experimental flow includes preparation and mounting of the slide, loading prime samples, assignment of the region of interests (ROIs), determination of operating angle, and data acquisition (Figure 17.2). In SPRi experiments, data generation is done in a real-time manner and generated data files are saved in proper format (as movies) and exported for further analysis.

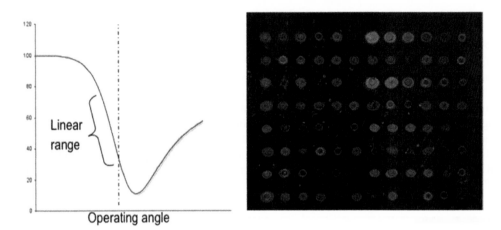

FIGURE 17.2   Selection of an operating angle and assignment of the region of interests (ROIs) in the SPRi experiment.

17.2 Different Applications of SPRi

SPRi-based biosensors are attractive choices for the detection of low-abundance analytes and biomolecular interactions in the HT manner (Ray et al., 2010; Reddy et al., 2012). Over the last decade, quite a few studies have used this label-free sensing platform for the direct detection of potential marker proteins in human serum and other biological samples (Table 17.1). In a study, Ladd et al. have applied SPRi in combination with an antibody array for the detection of cancer biomarkers in buffer solution and diluted human serum (10%) samples (Ladd et al., 2009). The authors have employed SPR imaging sensor with polarization contrast and detected activated leukocyte cell adhesion molecule/CD 166 (ALCAM) and transgelin-2 (TAGLN2) biomarkers, down to ng/mL concentration (limit of detection – 6 and 3 ng/mL for ALCAM and TAGLN2, respectively) devoid of any significant cross-reactivity. Such studies prove the potential of SPRi-based biosensors for the detection of biomarkers from complex biological samples.

TABLE 17.1

Various Applications of SPRi-Based Biosensors (Selected Examples)

Application	Studies (Reference)
(a) Study of biomolecular interactions	1. Protein-protein interactions (Natarajan et al., 2008)
	2. Interactions of GST-fusion proteins (Yuk et al., 2006)
	3. Structural changes with denaturation (Huang & Chen, 2006)
	4. Adsorption and desorption of multiple proteins, protein-polymer interactions (Hook et al., 2009)
	5. Antigen-antibody reactions (Xinglong et al., 2005)
(b) Detection of disease biomarkers and drug discovery	1. Cancer; activated leukocyte cell adhesion molecule/CD 166 (ALCAM) and transgelin-2 (Ladd et al., 2009), prostate specific antigen (PSA) (Malic et al., 2011)
	2. Auto-antibodies in sera of autoimmune patients (Lokate et al., 2007)
	3. Drug discovery applications (Olaru et al., 2015)
	4. Screening of cancer inhibitors (Basappa et al., 2011)

However, the sensitivity obtained with actual biological samples is found to be lower as compared to that obtained with buffer solutions. To increase the sensitivity of SPRi-based biosensors different nano-structured materials, particularly quantum dots and gold nanoparticles are introduced for signal amplification. Application of nanoparticles in SPRi can effectively increase the detection limit and target antigen to be detected at pM (picromolar) level, which is not possible to achieve with conventional SPRi settings.

17.3 Advantages and Disadvantages of SPRi

Advantages
The major advantage of SPRi over standard SPR is its multiplexing capabilities (ability to observe hundreds of reactions simultaneously). SPRi can be applied to perform interaction studies in a HT manner, which is not possible by conventional SPR. Additionally, other advantages of SPR-based biosensors are also applicable for SPRi:

- Label-free detection eliminates the requirement of any secondary reactants and long labeling process.
- Real-time measurements of biomolecular interaction and binding kinetics.
- High potential for multiplex analysis; so very well suited for high-throughput analysis.
- Both quantitative and qualitative measurements are possible.

Disadvantages
Due to its HT capabilities, SPRi is promising for clinical research, but there are many limitations as well, such as:

- Requires sophisticated instrumentation.
- Restricted to the choice of metal (gold/silver surfaces).
- Change in temperature affects refractive index and thereby the efficiency of the measurement.
- Non-specific interactions affect the signal.
- Heterogeneous sample surface affects sensitivity.
- Not very effective in handling complex biological samples.

17.4 Conclusions

SPRi-based biosensors have shown their potential to measure kinetic reactions of biomolecular interactions in HT manner. These techniques are very efficient for real-time measurement of disease-related proteins in the buffer *in vitro*. Some of the recent studies have testified the applicability of SPRi-based biosensors for direct detection of marker proteins in different biological fluids, including serum/plasma, saliva and urine. However, issues regarding sensitivity and specificity remain to be explored further when complex biological samples are concerned. Due to the requirement of sophisticated instrumentation and restriction to gold/silver surfaces, the detection cost of SPRi-based sensors is very high and not affordable for routine clinical diagnostics. If these basic limitations are circumvented successfully, SPRi could be one of the very attractive choices for HT proteomic research.

REFERENCES

Basappa, N. S., Elson, P., Golshayan, A.-R., Wood, L., Garcia, J. A., Dreicer, R., & Rini, B. I. (2011). The impact of tumor burden characteristics in patients with metastatic renal cell carcinoma treated with sunitinib. *Cancer, 117*(6), 1183–1189. https://doi.org/10.1002/cncr.25713

Hook, A. L., Thissen, H., & Voelcker, N. H. (2009). Surface plasmon resonance imaging of polymer microarrays to study protein–polymer interactions in high throughput. *Langmuir, 25*(16), 9173–9181. https://doi.org/10.1021/la900735n

Huang, H., & Chen, Y. (2006). Label-free reading of microarray-based proteins with high throughput surface plasmon resonance imaging. *Biosensors and Bioelectronics, 22*(5), 644–648. https://doi.org/10.1016/j.bios.2006.01.025

Ladd, J., Taylor, A. D., Piliarik, M., Homola, J., & Jiang, S. (2009). Label-free detection of cancer biomarker candidates using surface plasmon resonance imaging. *Analytical and Bioanalytical Chemistry, 393*(4), 1157–1163. https://doi.org/10.1007/s00216-008-2448-3

Lokate, A. M. C., Beusink, J. B., Besselink, G. A. J., Pruijn, G. J. M., & Schasfoort, R. B. M. (2007). Biomolecular interaction monitoring of autoantibodies by scanning surface plasmon resonance microarray imaging. *Journal of the American Chemical Society, 129*(45), 14013–14018. https://doi.org/10.1021/ja075103x

Malic, L., Sandros, M. G., & Tabrizian, M. (2011). Designed biointerface using near-infrared quantum dots for ultrasensitive surface plasmon resonance imaging biosensors. *Analytical Chemistry, 83*(13), 5222–5229. https://doi.org/10.1021/ac200465m

Natarajan, S., Katsamba, P. S., Miles, A., Eckman, J., Papalia, G. A., Rich, R. L., Gale, B. K., & Myszka, D. G. (2008). Continuous-flow microfluidic printing of proteins for array-based applications including surface plasmon resonance imaging. *Analytical Biochemistry, 373*(1), 141–146. https://doi.org/10.1016/j.ab.2007.07.035

Olaru, A., Bala, C., Jaffrezic-Renault, N., & Aboul-Enein, H. Y. (2015). Surface plasmon resonance (SPR) biosensors in pharmaceutical analysis. *Critical Reviews in Analytical Chemistry, 45*(2), 97–105. https://doi.org/10.1080/10408347.2014.881250

Ray, S., Mehta, G., & Srivastava, S. (2010). Label-free detection techniques for protein microarrays: Prospects, merits and challenges. *Proteomics, 10*(4), 731–748. https://doi.org/10.1002/pmic.200900458

Reddy, P. J., Sadhu, S., Ray, S., & Srivastava, S. (2012). Cancer biomarker detection by surface plasmon resonance biosensors. *Clinics in Laboratory Medicine, 32*(1), 47–72. https://doi.org/10.1016/j.cll.2011.11.002

Xinglong, Y., Dingxin, W., Xing, W., Xiang, D., Wei, L., & Xinsheng, Z. (2005). A surface plasmon resonance imaging interferometry for protein micro-array detection. *Sensors and Actuators B: Chemical, 108*(1–2), 765–771. https://doi.org/10.1016/j.snb.2004.12.089

Yuk, J. S., Kim, H.-S., Jung, J.-W., Jung, S.-H., Lee, S.-J., Kim, W. J., Han, J.-A., Kim, Y.-M., & Ha, K.-S. (2006). Analysis of protein interactions on protein arrays by a novel spectral surface plasmon resonance imaging. *Biosensors & Bioelectronics, 21*(8), 1521–1528. https://doi.org/10.1016/j.bios.2005.07.009

Exercises 17.1

1. Which of the following inherent properties of query molecules is measured in SPRi?
   a. Optical phase differences
   b. Alterations in conductivity
   c. Change in density
   d. Change in refractive index

2. Which of the following component(s) is present in SPRi but not in SPR?
   a. Prism
   b. Light source
   c. Imaging device (CCD camera)
   d. All of the above

3. SPRi can be used for which of the following application(s)?
   a. Study of biomolecular interactions
   b. Screening of inhibitors of tumor targets
   c. Detection of cancer biomarkers
   d. All of the above

4. Which of the following candidate(s) are used as signal amplifiers in SPRi?

    a. Quantum dots

    b. Gold nanoparticles

    c. Both of the above

    d. None of the above

5. In SPRi, intensity of the incident light and wavelength …?

    a. Remains constant

    b. Differs with time

    c. Differs with region of interest

    d. Intensity of the incident light remains fixed, but wavelength is variable

6. Which of the following statements is *not* true for SPRi?

    a. Both quantitative and qualitative measurements are possible

    b. Potential for multiplex analysis

    c. Real-time measurements

    d. Extremely effective in handling complex biological samples

7. Which one of the following can act as a reference surface for SPR assays?

    a. Unmodified surface

    b. Activated-deactivated surface

    c. Surface immobilized with dummy ligand

    d. All of the above

8. SPR imaging does not require labeling of the target molecule for detection.

    a. True

    b. False

9. In an SPR assay, the steps involved in an immobilization event occur in which of the following order?

    a. Immobilization, activation of the surface, deactivation

    b. Activation of the surface, immobilization, deactivation

    c. Deactivation, activation of the surface, immobilization

    d. Immobilization, activation of the surface, deactivation

10. Given below is the sensorgram obtained from a SPRi analysis. The concentrations of protein which was used in this experiment are 1, 10, and 100 nM. Your task is to identify which of the curves correspond to the given concentrations.

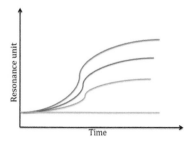

Answers

1. d
2. c
3. d
4. c
5. a
6. d
7. d
8. a
9. b
10. The orange curve/line represents the baseline, which was obtained before starting the assay. The y-axis on this graph represents the signal intensity and the signal intensity is directly related to the concentration of the protein used in this assay. Based on this one can infer that the blue, red, and green curves represent the curves obtained from the protein of concentrations 1, 10, and 100 nM, respectively.

18

Protein Interaction Analysis Using SPR and SPRi

Preamble

The study of protein-protein interactions is very important to understand the biological systems since these interactions play a central role in vital physiological pathways, cellular communications and host-pathogen interactions. Over the last few decades, quite a few popular approaches, including two-hybrid system, co-immunoprecipitation and affinity chromatography, have been routinely used for studying protein-protein interactions as well as the interaction of proteins with other biomolecules and small ligands. In recent years, protein and antibody microarrays and SPR-based sensing approaches have emerged for studying multiple protein-protein interactions simultaneously. SPR and SPRi-based biosensors are the most potential approaches for studying protein-protein interactions in a high-throughput manner. In this chapter, we will discuss the application of SPR and SPRi-based biosensors for studying protein-protein interactions.

Terminology

- **SPR:** Surface plasmon resonance (SPR) is a label-free method that measures variations in the refractive index of the dielectric layer adjoining to the sensor surface as a result of the adsorption or desorption of molecules.
- **SPRi:** SPR imaging (SPRi) is an advanced and high throughput platform of SPR, where the complete biochip surface is illuminated using a broad beam, monochromatic, polarized light, and a CCD camera is used for simultaneous capturing of reflected light from each spot.
- **Sensorgram:** Sensorgram is a graphical representation of SPR data, which indicates the binding response in reflection intensity versus time.
- **Protein-protein interaction:** Interaction/binding between two or more proteins, which often regulate the activity of the proteins. Interactions between proteins play crucial roles in vital physiological pathways, cellular communications and host-pathogen interactions.
- **Kinetics:** Parameters indicating the rates of forward and reverse reactions.
- **Affinity:** Affinity indicates the strength of an interaction, which is measured quantitatively by the dissociation constant.
- **Dissociation constant (K_D):** It is the ratio of on rate (K_a) and off rate (K_d), which reflects the strength of interaction.
- **On rate (K_a):** Also known as association rate and indicates the rate of forward reaction leading to the product formation.
- **Off rate (K_d):** Also known as dissociation rate and indicates the rate of reverse reaction leading to the dissociation of the complex.

18.1 Surface Plasmon Resonance and SPR Imaging: Comparative Analysis

Surface plasmon resonance (SPR) based sensing approaches depend on the measurement of change in the refractive index of the medium directly in contact with the sensor surface. In SPR measurement, the sensorgram indicates the changes in reflection intensity with respect to the incident angle when the target molecule binds to the sensor surface (Figure 18.1). Alteration in the refractive angle is directly proportional to the mass bound to the surface.

DOI: 10.1201/9781003098645-22

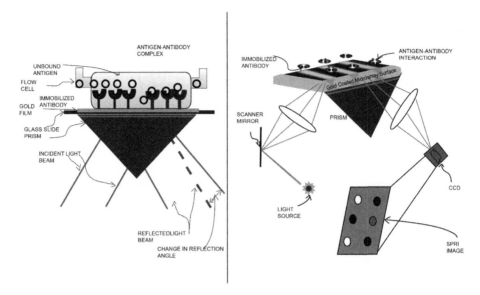

FIGURE 18.1 The basic working principle of (left) SPR and (right) SPRi for monitoring protein-protein interactions.

Although SPR and surface plasmon resonance imaging (SPRi) follow the same working principle for bio-sensing, there are the following major differences in SPRi instrumental set-ups.

- In SPRi, a CCD camera is used for instantaneous capturing of reflected light from each spot of the sensor surface, which allows instantaneous analysis of binding events on all spots.
- SPRi is more suitable for high throughput (HT) analysis compared to SPR.

Both SPR and SPRi have the following advantages regarding the analysis of protein-protein interactions:

- Allow label-free detection without the need for any secondary reactants and extensive labeling process.
- Real-time measurements.
- Provide information regarding binding kinetics (rates of association and dissociation).
- Suitable for studying multiple biomolecular interactions simultaneously.
- Provide both quantitative and qualitative measurements.

In this label-free detection method, one of the interaction partners is immobilized on the sensor surface and the second protein is allowed to pass through. The interaction between the interacting protein partners is monitored by measuring changes in the refractive angle, which is inversely proportional to the mass bound to the sensor surface.

18.2 Study of Protein-Protein Interactions Using SPR and SPRi

SPR-based sensing approaches have been successfully used in studying protein-protein and other biomolecular interactions (Table 18.1) (Huber & Mueller, 2006; Torreri et al., 2005). To analyze interactions between two proteins using SPR-based sensing approaches, one of the interaction partners is immobilized on the sensor surface while the other one is injected into the solution (Willander & Al-Hilli, 2009). The binding of the second partner on the sensor surface due to biomolecular interactions leads to change in the refractive angle, which is inversely proportional to the mass bound at the sensor surface. The quantity of bound materials is continuously measured as a function of time. After monitoring the interactions for a certain period, the buffer solution is changed to stop the interaction to monitor the dissociation of the complex formation (Figure 18.2).

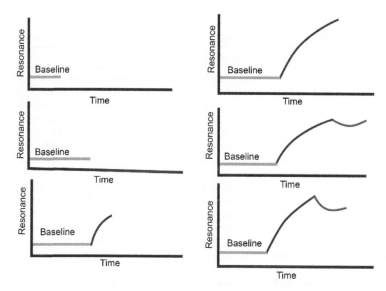

FIGURE 18.2 Determination of association and dissociation kinetics of biomolecular interaction using SPR.

Different steps are involved in biomolecular interaction analysis using the SPR sensor. The interaction between the molecules is measured by plotting the sensorgram as shown in panels. SPR provides real-time measurements of interactions; for that reason, the kinetics of interactions [on rate (K_a) and off rate (K_d)] can be calculated.

$$A + B \overset{K_a}{\underset{K_d}{\leftrightarrow}} AB.$$

The association rate constant of interaction can be calculated from the association period data if the concentration of interacting partners is known. To measure the dissociation rate constant(s) exponential decay to the dissociation data are used (Berggård et al., 2007). There are several published studies where SPR/SPRi has been used for measuring protein-protein or protein-small molecule interactions; few studies are described in tabulated format here (Table 18.1).

18.3 Data Processing and Analysis

Processing of SPR raw data before analysis is essential to obtain a superior interpretation of data generated during the association/dissociation phase of SPR-based analysis of protein-protein interactions. After obtaining raw data from the experimental process, Y-axis transformation is performed to fit the baselines of different time points and concentration variations (Figure 18.3). There are numerous commercially available software to fit curves and obtain apparent rate constants. More often the presence of background spot intensity significantly reduces artifacts and improves the Signal/Noise ratio and thereby the quality of data. To obtain sensorgram for kinetics analysis, subtraction of background spots is essential (Figure 18.4). Response from the reference surface and response of buffer injection are subtracted before analysis of SPR datasets. Considering a bidirectional interaction between two interacting partners, the value for K_D (ratio of K_{on}/K_{off}) is generally calculated using a nonlinear least-squares analysis (Oshannessy et al., 1993).

To obtain kinetic constants, various binding models like simple Langmuir binding models are applied. Controls are prepared using an empty surface for baseline checking. To determine the reaction kinetics between two interacting partners, one partner is immobilized and multiple concentrations of the second analyte are used and interactions are analyzed. The quality of the obtained data can be evaluated by verifying the K_{on} and K_{off} values. Kinetics of binding is very crucial while selecting the drug targets and screening of potential drug molecules. In SPR and SPRi, the major advantage is that multiple interactions can be monitored simultaneously, which allows detection of interactions of twol interacting partners using a

TABLE 18.1

Application of SPR and SPRi for Studying Protein-Protein or Protein-Small Molecule Interactions

Study	Detection Technique	Description
Antigen-antibody (IL6 and anti-IL6) interactions (Rispens et al., 2011)	SPR	Interaction between anti-interleukin 6 (IL6) antibodies, anti-IL6.16 and anti-IL6.8 with interleukin 6 (chips coated with either anti-IL6.16 or anti-IL6.8). Timepoint analysis was performed using different concentrations of IL6 and two antibodies (anti-IL6.16 and anti-IL6.8). Higher on-rate and smaller off-rate were found for anti-IL6.8 than anti-IL6.16, indicating a higher affinity of anti-IL6.8 for interleukin 6.
Studying association or dissociation kinetics of diverse Fab and hK1 interaction (Wassaf et al., 2006)	SPR in combination with microarray	The overall aim of this study was the detection of antibodies that bind at the active site of human tissue kallikrein 1 (hK1) and consequently inhibit the protease activity of hK1. Simultaneous analysis of kinetic constants (k_{on} and k_{off}) for 96 different F_{ab} fragments using array format. F_{abs} were categorized on basis of their capacity to recognize an apparent active site epitope. Immobilization of Fab was performed using specific capture surfaces (anti-cMyc or protein A).
IgG and Protein A interaction (Natarajan et al., 2008)	SPRi in combination with continuous-flow microfluidics (CFM)	This study demonstrated that coupling of continuous-flow microfluidics with SPRi could improve the printing process. It allows immobilization of purified proteins on the gold surface at a very low concentration. Low sample consumption and multiplexing capability are major advantages of this combined technology.
Clinically related protein-peptide interactions (Cherif et al., 2006)	SPRi	Interactions of three peptides (C 20–40, C 131–150, and Ova 273–288) with rabbit anti-C 20–40 and anti-C 131–150 immune sera (1:100 diluted). Detection of such weak clinically relevant interactions is promising for clinical research.
Antibody-antigen binding (Nogues et al., 2010)	SPRi in combination with peptide microarrays	65kDa isoform of human glutamate decarboxylase (GAD65) and a human monoclonal antibody. Specific bindings of Rac1 and RhoA antibodies to their antigens immobilized on protein arrays were monitored by spectral SPR imaging. Kinetic parameters of the interaction were measured more efficiently than ELISA/RIA methods.
Interactions of GST-fusion proteins with their antibodies (Yuk et al., 2006)	SPRi (Wavelength interrogation-based self-constructed with Kretschmann-Raether geometry)	Interactions of glutathione S-transferase-fusion proteins with their antibodies; anti-Rac1 and anti-RhoA to Rac1 and RhoA have been studied. Quartz tungsten halogen lamp was applied as a light source. Protein arrays were prepared by immobilizing glutathione S-transferase (GST) fusion proteins on the glutathione surfaces. Protein arrays were analyzed by two-dimensional images.

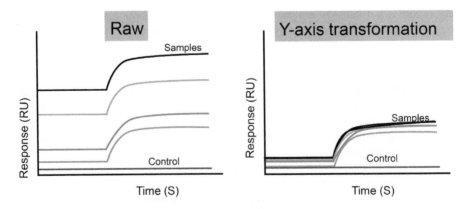

FIGURE 18.3 Y-axis transformation of SPR sensorgram to fit data.

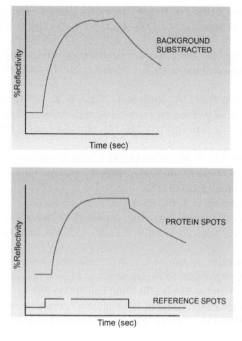

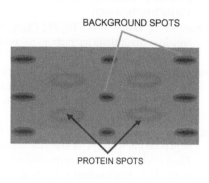

FIGURE 18.4 Subtraction of background reference spot intensity in SPR data analysis.

different concentration of ligands. The interaction of multiple analytes to the same ligand can also be monitored and binding affinities can be compared. Even if multiple molecules have the same affinity, they may have different K_{on} and K_{off}, and specific candidates can be selected according to the need of the investigator.

18.4 Conclusions

Although SPR and SPRi-based biosensors are suitable for diverse types of applications, maximum use of these label-free sensing approaches is found for the analysis of biomolecular interactions, since it provides real-time information regarding equilibrium binding constants, kinetic rate constants and thermodynamic parameters. Over the last decade, a large number of studies have proved the utility of SPR and SPRi-based biosensors for protein-protein and protein-small molecule interactions. Additionally, interactions of proteins with other types of biomolecules including protein-carbohydrate interactions (Linman et al., 2008; Yuk et al., 2006), DNA-protein interaction (Stockley & Persson, 2009) etc., are also vividly studied by SPR and SPRi. Despite multiple advantages, SPR-based approaches have not yet gained extreme popularity in routine clinical use, mostly due to the detection cost associated with the requirement of sophisticated instrumentation and restriction to gold/silver surfaces. To extend the applications of SPR-based techniques, diverse amalgamated technological approaches have been developed such as SPR-MS, electrochemical SPR, and surface plasmon fluorescence spectroscopy, which have successfully circumvented some of the basic limitations associated with SPR.

REFERENCES

Berggård, T., Linse, S., & James, P. (2007). Methods for the detection and analysis of protein–protein interactions. *PROTEOMICS*, *7*(16), 2833–2842. https://doi.org/10.1002/pmic.200700131

Cherif, B., Roget, A., Villiers, C. L., Calemczuk, R., Leroy, V., Marche, P. N., Livache, T., & Villiers, M.-B. (2006). Clinically related protein–peptide interactions monitored in real time on novel peptide chips by surface plasmon resonance imaging. *Clinical Chemistry*, *52*(2), 255–262. https://doi.org/10.1373/clinchem.2005.058727

Huber, W., & Mueller, F. (2006). Biomolecular interaction analysis in drug discovery using surface plasmon resonance technology. *Current Pharmaceutical Design*, *12*(31), 3999–4021. https://doi.org/10.2174/138161206778743600

Linman, M. J., Taylor, J. D., Yu, H., Chen, X., & Cheng, Q. (2008). Surface plasmon resonance study of protein–carbohydrate interactions using biotinylated sialosides. *Analytical Chemistry*, *80*(11), 4007–4013. https://doi.org/10.1021/ac702566e

Natarajan, S., Katsamba, P. S., Miles, A., Eckman, J., Papalia, G. A., Rich, R. L., Gale, B. K., & Myszka, D. G. (2008). Continuous-flow microfluidic printing of proteins for array-based applications including surface plasmon resonance imaging. *Analytical Biochemistry*, *373*(1), 141–146. https://doi.org/10.1016/j.ab.2007.07.035

Nogues, C., Leh, H., Langendorf, C. G., Law, R. H. P., Buckle, A. M., & Buckle, M. (2010). Characterisation of peptide microarrays for studying antibody-antigen binding using surface plasmon resonance imagery. *PLoS ONE*, *5*(8), e12152. https://doi.org/10.1371/journal.pone.0012152

Oshannessy, D. J., Brighamburke, M., Soneson, K. K., Hensley, P., & Brooks, I. (1993). Determination of rate and equilibrium binding constants for macromolecular interactions using surface plasmon resonance: Use of nonlinear least squares analysis methods. *Analytical Biochemistry*, *212*(2), 457–468. https://doi.org/10.1006/abio.1993.1355

Rispens, T., Te Velthuis, H., Hemker, P., Speijer, H., Hermens, W., & Aarden, L. (2011). Label-free assessment of high-affinity antibody-antigen binding constants. Comparison of bioassay, SPR, and PEIA-ellipsometry. *Journal of Immunological Methods*, *365*(1–2), 50–57. https://doi.org/10.1016/j.jim.2010.11.010

Stockley, P. G., & Persson, B. (2009). Surface plasmon resonance assays of DNA-protein interactions. In B. Leblanc & T. Moss (Eds.), *DNA-Protein Interactions* (Vol. 543, pp. 653–669). Humana Press. https://doi.org/10.1007/978-1-60327-015-1_38

Torreri, P., Ceccarini, M., Macioce, P., & Petrucci, T. C. (2005). Biomolecular interactions by surface plasmon resonance technology. *Annali Dell'Istituto Superiore Di Sanita*, *41*(4), 437–441.

Wassaf, D., Kuang, G., Kopacz, K., Wu, Q.-L., Nguyen, Q., Toews, M., Cosic, J., Jacques, J., Wiltshire, S., Lambert, J., Pazmany, C. C., Hogan, S., Ladner, R. C., Nixon, A. E., & Sexton, D. J. (2006). High-throughput affinity ranking of antibodies using surface plasmon resonance microarrays. *Analytical Biochemistry*, *351*(2), 241–253. https://doi.org/10.1016/j.ab.2006.01.043

Willander, M., & Al-Hilli, S. (2009). Analysis of biomolecules using surface plasmons. *Methods in Molecular Biology (Clifton, N.J.)*, *544*, 201–229. https://doi.org/10.1007/978-1-59745-483-4_14

Yuk, J. S., Kim, H.-S., Jung, J.-W., Jung, S.-H., Lee, S.-J., Kim, W. J., Han, J.-A., Kim, Y.-M., & Ha, K.-S. (2006). Analysis of protein interactions on protein arrays by a novel spectral surface plasmon resonance imaging. *Biosensors & Bioelectronics*, *21*(8), 1521–1528. https://doi.org/10.1016/j.bios.2005.07.009

Exercises 18.1

1. In SPRi, what is the processing of raw data shown in the figure below from left to right called?

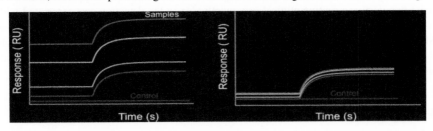

    a. Blank subtraction

    b. Y-axis transformation

    c. X-axis transformation

    d. None of the above

2. What is the most common application of SPR and SPRi?
   a. Study of biomolecular interactions
   b. Discovery of disease biomarkers
   c. Determination of protein structures
   d. Analysis of post-translational modifications (PTMs)

3. Which of the following technique(s) is used for studying protein-protein interactions?
   a. Yeast two-hybrid system
   b. SPR and SPRi
   c. Protein and antibody arrays
   d. All of the above

4. SPR and SPRi can provide information on which of the following regarding protein-protein interactions?
   a. Rates of association and dissociation
   b. Affinities
   c. Kinetics
   d. All of the above

5. Which of the following statements is true regarding SPR?
   a. Changes in the refractive angle is directly proportional to the mass bound at the surface
   b. Changes in the refractive angle is inversely proportional to the mass bound at the surface
   c. Changes in the refractive angle do not depend on the mass bound at the surface
   d. Changes in the refractive angle is dependent of the chemical nature of the sample being analyzed

6. Is there a difference between response bound and response final?
   a. No, they are the same
   b. Yes, response bound represents the amount of ligand covalently bound to the surface
   c. Yes, response final represents the amount of ligand covalently bound to the surface

7. Which of the following statement(s) is true about the SPR-MS application?
   a. Can identify the specific protein partner from a heterogeneous analyte mixture
   b. Desired interacting protein can be collected in vials for further MS analysis
   c. The same binding assay is run multiple times to collect higher amount of protein
   d. All of the above

8. Is it possible for two different biomolecules to have similar affinities but different kinetics?
   a. Yes
   b. No

9. Specificity analysis of the molecules on SPR means?
   a. Yes/no screening
   b. Immobilization of lot of molecules simultaneously on a chip
   c. Binding analysis of drug compounds only
   d. None of the above

10. SPR-based assays cannot be used to reveal structural and conformational changes in proteins.
    a. True
    b. False

11. Given below are the graphs obtained from an SPR experiment. In this experiment, two different proteins are compared. Comment on the interaction of these proteins with the immobilized protein and dissociation constants based on the graphs given below.

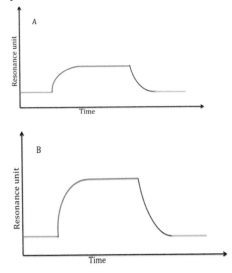

Answers

    1. b
    2. a
    3. d
    4. d
    5. a
    6. c
    7. d
    8. a
    9. a
    10. a
    11. The signal intensity in SPR is directly proportional to the amount of the protein binding to the surface. Based on the graphs, it seems that the amount of protein B bound to the surface is more than the protein A. Hence one can observe greater signal intensity (shown as red curve in the graphs) when protein B is used. To compare the dissociation constants K_D one need Ka (on rate) and Kd (off rate) information, which is not provided here.

Module V

Advancements in Proteomics

19

Proteomics for Translational Research

Preamble

Proteomics is growing rapidly as a promising platform to translate the knowledge obtained from basic research into diverse clinical applications. Proteomics in clinical research has successfully accelerated the identification of potential diagnostic and prognostic biomarkers and novel drug and vaccine targets for improvement of diagnostics and therapeutics. Although the major focus of proteomics research has focused on oncology, different potential biomarkers for infectious diseases, autoimmune diseases, and cardiovascular diseases have also been identified and potential drug/vaccine candidates have been evaluated. Apart from the clinical fields, proteomics research has also been implemented in industrial applications including crop improvement, sanitation assessment, allergens, toxins and food-borne pathogen detection. In this chapter, we will discuss proteomics for translational research in clinical fields and the prospects and challenges associated with the translation of proteomics knowledge in clinical applications.

19.1 Translation of Proteomics in Clinical Research: Prospects and Challenges

Different emerging proteomic techniques including gel-based profiling, quantitative mass spectrometry and array-based high-throughput proteomics have shown considerable potential in differential proteomic expression analysis for the identification of potential biomarkers with promising diagnostic and prognostic predictions (Paulo et al., 2012; Ray, Reddy, Jain, et al., 2011). Different nanoproteomics technologies are found to be very efficient in the targeted approach for selective detection of low-abundant target protein markers present in complex biological fluids (Ray, Reddy, Choudhary, et al., 2011). Additionally, recent studies have suggested the potential of label-free proteomics for the screening of probable drug molecules. Hence, the translation of proteomics research in clinical fields encompasses (Figure 19.1) the following:

- Identification of novel diagnostic and prognostic marker proteins for cancer and other human diseases.
- Ultra-sensitive detection of marker proteins in biological fluids in a targeted approach.
- Identification of novel therapeutic targets (drugs and vaccines).
- Individualized healthcare and personalized proteomics for medicine/therapy.
- Identification of new drug molecules and elucidation of the mechanism of action of known drugs (Figure 19.2).

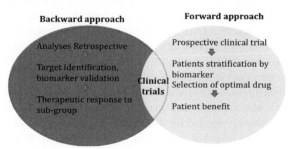

FIGURE 19.1 Identification of potential biomarkers by two different approaches – prospective and retrospective.

DOI: 10.1201/9781003098645-24

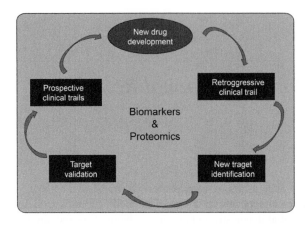

FIGURE 19.2   Depiction of the process of drug development and decision making by utilizing proteomics.

Despite the immense applicability of different proteomics techniques, translation of knowledge obtained from basic proteomics research into direct clinical applications is challenging, indicating a huge gap between the "bench-side" research and "bed-side" implications. Due to the dynamic nature of proteins and complexity of proteomes as well as extreme variations among individuals, often it becomes difficult to translate the findings obtained under lab environment effectively in actual applications. There are the some basic limitations associated with translational proteomic research as below.

- The field is developing and quite a few approaches are still at the proof-of-principle level.
- There is no protein amplification method (like PCR used for gene amplification).
- The fragile character of proteins and difficulties in extraction and isolation of proteins from specific organelles.
- Presence of various isoforms of a single protein.
- Huge variations in proteome with time within the same individual.
- Variations among the individuals of same or different populations; difficult to establish gold-standard biomarkers applicable for all populations.
- The broad dynamic range of protein concentrations and complexity of biological fluids.

19.2 Proteomics in Diagnostics: Biomarker Discovery

Biomarkers are indicator biomolecules that assist in detecting diseased conditions at an early stage, discriminate between different diseases, and useful for monitoring the progression/severity of a disease. Proteomics is a useful platform for comprehensive analysis of protein expression levels in control and diseased conditions and can indicate the alterations in host proteome due to external infections or unhealthy conditions. Therefore, over the last decade, different researchers have adopted diverse proteome profiling techniques for the identification of marker proteins in biological samples, including serum/plasma, urine, saliva, tissue, cerebrospinal fluid (CSF), etc., and reported a plethora of potential proteins that can serve as classifier molecules for accurate discrimination of disease states from healthy normal, as well as can differentiate among multiple diseases (Anderson, 2010; Drake et al., 2005; Hu et al., 2006; Pisitkun et al., 2006). Gel-based proteomic technologies, particularly, classical two-dimensional electrophoresis (2DE) and 2D-DIGE have been massively applied for differential proteome profiling of diseased and normal samples. In recent years, quantitative MS techniques including iTRAQ, ICAT, MRM, label-free MS approaches and microarray-based approaches, particularly reverse-phase protein microarrays, gained popularity for the identification of protein markers for human diseases. Diagnostic markers help to detect disease at an early stage of development; while prognostic biomarkers assist in discriminating among diseased states and help to differentiate severe and non-severe forms of diseases (J. Lee et al., 2011). In proteomic analysis, workflow for biomarker discovery, generally in the initial discovery phase small-/moderate-sized populations are screened for identification of potential candidates, while in the validation phase, identified markers are tested using bigger clinical cohorts (Figure 19.3).

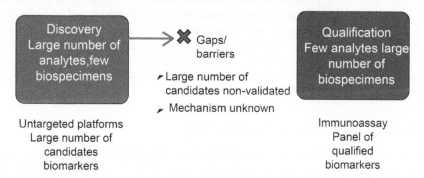

FIGURE 19.3 Representation of a typical proteomics research pipeline from discovery to qualification.

Once the possible candidates with an ability of discrimination between control and diseased states are identified, specificity, sensitivity and accuracy of the marker proteins are evaluated in multiple levels of trials. Often a combination of a panel of marker proteins is required rather than a single marker for accurate detection of any complex disease/disorder, which has similar/overlapping manifestations with other diseases (Figure 19.4). According to a 2008 report of the US Food and Drug Administration (FDA), protein-based assays for 109 unique protein targets in plasma or serum have been approved for further analysis (Anderson, 2010). Among the different approved markers, vascular endothelial growth factor (VEGF), haptoglobin, β-2 microglobulin, α-fetoprotein, carcinoembryonic antigen (CEA) and cancer antigen 125 (CA 125) are the most promising candidates for cancer diagnosis (Anderson, 2010). Different promising proteomic techniques have reported potential biomarkers for the early detection of cancers and other different types of infectious and non-infectious diseases. Unfortunately, a majority of them have not translated into practical clinical applications. The foremost impediments for the establishment of gold-standard biomarkers are associated with extreme biological variations and non-specificity of the identified marker for any definite diseases.

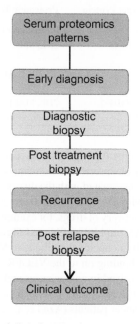

FIGURE 19.4 Flowchart shows the utility of clinical proteomics.

19.3 Proteomics in Therapeutics

Proteomics deals with the study of the entire proteome present in the specific cells/tissues at a given period. Proteins play a major role in various metabolic processes. Alterations in the protein levels/activity may disrupt the normal physiological conditions in the organisms and give rise to diseases (Cooper & Carucci, 2004). By using the proteomics approach, we can identify changes in various proteins and metabolic pathways in the organism under pathological conditions. By targeting the responsible agent for the disease, we can cure the disease. With the latest developments in the field of proteomics, investments have been increased in this field for the identification of new drug targets and the development of vaccines before the onset of the disease.

19.3.1 Identification of Drug and Vaccine Targets

Nowadays, proteomics is widely used in the identification of drug targets in various cancers and infectious diseases (Cooper & Carucci, 2004; Zwick et al., 2002) (Table 19.1). In the case of cancer, various kinases that are involved in the increased proliferation of cells are activated. These kinases are the proteins of interest for cancer drug development. Some of the drugs, which are targeted against the kinases are under the evaluation stage prior to their use in the clinical application (Traxler et al., 2001). In the case of malaria (*falciparum* malaria), cysteine proteases are good targets for drug development as they are involved in various metabolic pathways in the malarial parasite (B. J. Lee et al., 2003). In conventional drug development methods, drugs have been developed that act at the upstream levels resulting in the inhibition of the upstream products and non-formation of downstream products. However, this kind of approach often leads to side effects (toxic effects) when inhibitor/drug concentration is high. If we can develop drugs against a few target proteins involved in a metabolic process responsible for a particular disease, it will be better than targeting a single protein. In multi-protein targeting, the required amount of a drug will be minimal, which reduces the toxic side effects by preventing the accumulation of a particular drug (Elnaggar et al., 2012). Proteomics helps in the identification and characterization of various proteins involved in the metabolism based on which drugs can be designed against several proteins.

TABLE 19.1

List of Potential Drug Targets Identified in Various Diseases Using Proteomic Approaches

Disease	Potential Drug Target Protein	Proteomic Techniques	Reference
Prostate cancer	Phosphorylated Akt	Reverse-phase arrays	(Paweletz et al., 2001)
Falciparum malaria	Plasmepsins, plasmepsins-like enzymes	2DE, MS	(Cooper & Carucci, 2004)
Sepsis-induced acute renal failure	Merpin-1-α	2D-DIGE, MALDI-MS	(Holly et al., 2006)
Tuberculosis (Drug-resistant)	Rv1446c, Rv3028c, Rv0491, Rv2971, and Rv2145	2DE, MALDI-TOF-MS (Jiang et al., 2007)	(Jiang et al., 2007)

Additionally, the expression of proteins (up-regulation or down-regulation) in various pathological conditions can be studied by using various proteomic techniques. Up-regulation of the proteins associated with a metabolic pathway may be associated with the disease condition. In such conditions, the up-regulated proteins could be targeted, which may be potential targets for curing the disease. Such kinds of strategies are being explored for the discovery of drugs, for example, bevacizumab (monoclonal antibody to VEGF) along with sorafenib, a RAF (rapidly accelerated fibrosarcoma) kinase inhibitor is used for the treatment of cancers. Bevacizumab (monoclonal antibody to VEGF) sequesters the VEGF molecule and decreases its availability for their binding on to the VEGFR. On the other hand, Sorafenib inhibits the activity of RAF kinase which is involved in angiogenesis and activated in response to VEGF. Thus, inhibiting the VEGF signaling at two different points results in the reduction of angiogenesis. This treatment showed good results in ovarian cancer patients (Azad et al., 2008). The same strategy is being used in various cancer diseases for drug development. Quantitative proteomic techniques (iTRAQ, SELDI-TOF, ICAT, etc.) made it possible to identify the proteomic alterations in a large number of patients. The cell surface proteins of the parasites (in the case of infectious diseases) and the cancer cells (undergone alteration because of mutations) will be better targets for the development of antibody-based drugs as they are unique, the drugs developed against them act effectively and specifically on the parasite/cancer cells. However, such kind of drugs are not effective in the case of intracellular parasites.

Further, vaccines are used for the prevention of several diseases. Proteomics plays a major role in the identification of proper target molecules for the development of vaccines (Table 19.2). In the case of infectious diseases, the development of vaccines will be based on the identification of the cell surface proteins of the pathogen and the specific protein expression at the time of infection (Walters & Mobley, 2010). Servin et al. studied the cell surface antigens to develop vaccines against *Streptococcus pyogenes*. Cell surface antigens play a major role in the bacterial interaction with the host system and they are available for the host immune system to interact with and generate memory against the parasite (Severin et al., 2007). Proteomics-based approaches have been followed for the development of vaccines against various pathogens like *Staphylococcus aureus*, *Schistosoma mansoni*, *Francisella tularensis*, etc. (Eyles et al., 2007; Glowalla et al., 2009).

TABLE 19.2

Application of Proteomics for the Discovery of Potential Vaccine Targets

Microbe/Disease	Potential Target Protein (for Vaccine Development)	Proteomic Techniques	Reference
Helicobacter pylori	Le^b-binding adhesin, other membrane proteins	2D-LPE, MALDI-TOF-MS	(Nilsson et al., 2000)
Trypanosoma cruzi	TS/gp85, mucins and other membrane proteins	Multidimensional liquid chromatography, MALDI-TOF-MS	(Nakayasu et al., 2012)
Streptpcoccus sps.	spy0416	Nano-LC/MS/MS	(Rodríguez-Ortega et al., 2006)
Klebsiella pneumoniae	OmpA, OmpK36, OmpK17, FepA, OmpW, Colicin I receptor	2-DE, Immunoblotting, MALDI-TOF-MS	(Kurupati et al., 2006)
Streptococcus suis serotype 2	Muramidase-released protein, surface protein SP1, and glyceraldehyde-3-phosphate dehydrogenase (GapdH)	2-DE, Immunoblotting, MALDI-TOF-MS	(Zhang et al., 2008)

19.3.2 Screening of Potential Drug Molecules

With the advances in the field of proteomics, it is possible to screen a large number of drug molecules in a short time by using various labeled/label-free proteomic techniques. Proteomic techniques like affinity chromatography, mass spectrometry, microarrays and surface plasmon resonance (SPR), revolutionized the drug discovery by high-throughput screening of the drug targets (Table 19.1). In the affinity chromatography method of screening, the drug molecules which are assumed to bind with some of the proteins from the cell lysates are immobilized onto a column and the protein sample/cell lysate is allowed to pass over the immobilized ligand. After passing the protein sample, the column is washed with wash buffers to remove the unbound proteins. Then the proteins that are bound to ligand (drug) molecule are eluted out by altering the pH of the buffer (elution buffer) followed by their separation on SDS-PAGE. The protein bands observed on the polyacrylamide gel are subjected to mass spectrometry for the identification of the protein molecules, which are interacting with the drug molecule (Sleno & Emili, 2008).

Reverse-phase microarrays and tissue microarrays are used to screen proteomic alterations specific to a disease in a large number of patients. In this method, the patient's tumor tissue lysates (in the case of cancer) and the normal tissue lysates are spotted on the nitrocellulose slide separately. Antibody against the target protein is applied onto the slides and the expression levels are detected by using a secondary antibody tagged with an enzyme or a fluorescent molecule. The drug molecules can be developed and targeted against these target proteins. The drug molecules, which bind only to the tumor tissue lysates, could be further screened for their utility in clinical applications. Whole-body arrays contain the tissue lysates from the various tissues of the body, which are useful for the identification of the probable drug cross-reactive sites from various parts of the body (Petricoin & Liotta, 2003).

Label-free technique, SPR is used to study various protein-ligand interactions which are weak. SPR uses the total internal reflection as its principle for the identification of molecular interactions. The proteins of interest or the tissue lysates from the cancer cells/diseased tissue are immobilized onto a thin glass plate coated with a metallic film (generally gold coating). Then the target drug is allowed to pass over the tissue lysate or specific proteins on the SPR chip. If the proteins on the chip bind to the drug molecules, it results in an increase in mass at that point on the chip, which can be detected by the changes in the refractive index of the medium. A large number of drug molecules screening have been performed using such technology (Jönsson et al., 1991).

Apart from the clinical fields, proteomics research has also been implemented in industrial applications including crop improvement, sanitation assessment, allergens, toxins and food-borne pathogen detection and others as given in Table 19.3.

TABLE 19.3

Application of Proteomics in Different Translational Research (A Few Illustrative Examples)

Field	Application	Proteomic Techniques	Reference
Food industry	Identification of food allergens	2-DE, western blotting	(Akagawa et al., 2007)
Agriculture	Study of proteomic alterations in wheat grain in response to heat stress	2-DE, MALDI-TOF-MS	(Majoul et al., 2004)
Oil industry	Development of high oil yielding *Jatropha curcas* plants	2-DE, LTQ-ESI-MS/MS	(Yang et al., 2009)
Sports	To detect the doping agents	LC-MS/MS	(Kay & Creaser, 2010)
Pharmaceutical industry	To detect the purity of the drugs	MS	(Monsarrat et al., 1980)

19.4 Personalized Proteomics to Precision Medicine/Therapy

Personalized medicine and therapy are the major focuses of healthcare in recent years (Jain, 2004). Individualized therapy and personalized omics analysis are gaining popularity due to the extreme variations among the individuals and each person's unique response toward drug treatment and diseased states. Proteomics aids in understanding the biological and molecular functions of proteins, their involvements in different physiological pathways and interaction networks. It has tremendous implications for therapeutic interventions and identifying body responses during and post-drug treatment. After obtaining success with pharmacogenetic and pharmacogenomic approaches, pharmacoproteomics is also emerging effectively, since proteomics-based characterization of diseases and clinical groups are more appreciable for individual's preventive care or drug therapy (Ray, Reddy, Jain, et al., 2011; Reddy et al., 2011). Characterization of personalized omics including genomics, transcriptomics, proteomics and metabolomics is highly effective for the establishment of disease-oriented medicine and identification of at-risk personals and families with susceptibility to hereditary cancer and other fetal disorders (Chen & Snyder, 2013).

The major purpose of personalized omics and individualized therapy include:

- Identification of gene and protein-level variations among individuals.
- Establishment of personal omics profile.
- Analysis of individual's response pre/during/post-disease state.
- Development of patient-tailor therapy.
- Understanding disease pathobiology and early detection of diseased conditions.

However, the field of pharmacoproteomics and personalized proteomics is presently at an early stage of development. The fundamental improvement in proteomics research, data mining and systems biology approaches are required for making this discipline more effective and reliable for clinical applications.

19.5 Conclusions

The ultimate aspiration of clinical proteomics is to translate the findings of basic research into direct practical clinical applications through the development of processes/products, which can effectively improve diagnostics and therapeutics and in turn, provide beneficial outcomes for mankind. Despite the promising outcomes in proof-of-principle level experiments, proteomics is far away from the routine implications for disease diagnosis and treatment. In order to obtain a comprehensive picture of the living system, it is essential to perform organized studies by combining the proteomics findings with other omics level research, which can effectively unravel mechanistic insights of biological networks for the identification of novel diagnostic and therapeutic targets. High-throughput proteomic technologies competent for screening a large number of analytes rapidly are very promising for the identification of new drug molecules and diagnostic indicators. The eventual success of proteomics in translational research would depend on the worldwide initiatives for sharing of scientific findings among different research groups through the development of data repositories, correlation of new research findings with existing knowledge and circumvention of the fundamental limitations associated with proteome level research. To this end, the establishment of the Human Proteome Organization (HUPO), Human Proteome Project (HPP) and other proteome projects are phenomenal achievements for the translation of proteome research into direct "bed-side" applications.

REFERENCES

Akagawa, M., Handoyo, T., Ishii, T., Kumazawa, S., Morita, N., & Suyama, K. (2007). Proteomic analysis of wheat flour allergens. *Journal of Agricultural and Food Chemistry, 55*(17), 6863–6870. https://doi.org/10.1021/jf070843a

Anderson, N. L. (2010). The clinical plasma proteome: A survey of clinical assays for proteins in plasma and serum. *Clinical Chemistry, 56*(2), 177–185. https://doi.org/10.1373/clinchem.2009.126706

Azad, N. S., Posadas, E. M., Kwitkowski, V. E., Steinberg, S. M., Jain, L., Annunziata, C. M., Minasian, L., Sarosy, G., Kotz, H. L., Premkumar, A., Cao, L., McNally, D., Chow, C., Chen, H. X., Wright, J. J., Figg, W. D., & Kohn, E. C. (2008). Combination targeted therapy with sorafenib and bevacizumab results in enhanced toxicity and antitumor activity. *Journal of Clinical Oncology: Official Journal of the American Society of Clinical Oncology, 26*(22), 3709–3714. https://doi.org/10.1200/JCO.2007.10.8332

Chen, R., & Snyder, M. (2013). Promise of personalized omics to precision medicine. *Wiley Interdisciplinary Reviews. Systems Biology and Medicine, 5*(1), 73–82. https://doi.org/10.1002/wsbm.1198

Cooper, R. A., & Carucci, D. J. (2004). Proteomic approaches to studying drug targets and resistance in *Plasmodium*. *Current Drug Targets. Infectious Disorders, 4*(1), 41–51. https://doi.org/10.2174/1568005043480989

Drake, R. R., Cazares, L. H., Semmes, O. J., & Wadsworth, J. T. (2005). Serum, salivary and tissue proteomics for discovery of biomarkers for head and neck cancers. *Expert Review of Molecular Diagnostics, 5*(1), 93–100. https://doi.org/10.1586/14737159.5.1.93

Elnaggar, M., Giovannetti, E., & Peters, G. J. (2012). Molecular targets of gemcitabine action: Rationale for development of novel drugs and drug combinations. *Current Pharmaceutical Design, 18*(19), 2811–2829. https://doi.org/10.2174/138161212800626175

Eyles, J. E., Unal, B., Hartley, M. G., Newstead, S. L., Flick-Smith, H., Prior, J. L., Oyston, P. C. F., Randall, A., Mu, Y., Hirst, S., Molina, D. M., Davies, D. H., Milne, T., Griffin, K. F., Baldi, P., Titball, R. W., & Felgner, P. L. (2007). Immunodominant *Francisella tularensis* antigens identified using proteome microarray. ©Crown Copyright 2007 Dstl. *PROTEOMICS, 7*(13), 2172–2183. https://doi.org/10.1002/pmic.200600985

Glowalla, E., Tosetti, B., Krönke, M., & Krut, O. (2009). Proteomics-based identification of anchorless cell wall proteins as vaccine candidates against *Staphylococcus aureus*. *Infection and Immunity, 77*(7), 2719–2729. https://doi.org/10.1128/IAI.00617-08

Holly, M. K., Dear, J. W., Hu, X., Schechter, A. N., Gladwin, M. T., Hewitt, S. M., Yuen, P. S. T., & Star, R. A. (2006). Biomarker and drug-target discovery using proteomics in a new rat model of sepsis-induced acute renal failure. *Kidney International, 70*(3), 496–506. https://doi.org/10.1038/sj.ki.5001575

Hu, S., Loo, J. A., & Wong, D. T. (2006). Human body fluid proteome analysis. *PROTEOMICS, 6*(23), 6326–6353. https://doi.org/10.1002/pmic.200600284

Jain, K. K. (2004). Role of pharmacoproteomics in the development of personalized medicine. *Pharmacogenomics*, *5*(3), 331–336. https://doi.org/10.1517/phgs.5.3.331.29830

Jiang, X., Gao, F., Zhang, W., Hu, Z., & Wang, H. (2007). Comparison of the proteomes of isoniazid-resistant *Mycobacterium tuberculosis* strains and isoniazid-susceptible strains. *Zhonghua Jie He He Hu Xi Za Zhi = Zhonghua Jiehe He Huxi Zazhi = Chinese Journal of Tuberculosis and Respiratory Diseases*, *30*(6), 427–431.

Jönsson, U., Fägerstam, L., Ivarsson, B., Johnsson, B., Karlsson, R., Lundh, K., Löfås, S., Persson, B., Roos, H., & Rönnberg, I. (1991). Real-time biospecific interaction analysis using surface plasmon resonance and a sensor chip technology. *BioTechniques*, *11*(5), 620–627.

Kay, R. G., & Creaser, C. S. (2010). Application of mass spectrometry-based proteomics techniques for the detection of protein doping in sports. *Expert Review of Proteomics*, *7*(2), 185–188. https://doi.org/10.1586/epr.10.11

Kurupati, P., Teh, B. K., Kumarasinghe, G., & Poh, C. L. (2006). Identification of vaccine candidate antigens of an ESBL producing *Klebsiella pneumoniae* clinical strain by immunoproteome analysis. *Proteomics*, *6*(3), 836–844. https://doi.org/10.1002/pmic.200500214

Lee, B. J., Singh, A., Chiang, P., Kemp, S. J., Goldman, E. A., Weinhouse, M. I., Vlasuk, G. P., & Rosenthal, P. J. (2003). Antimalarial activities of novel synthetic cysteine protease inhibitors. *Antimicrobial Agents and Chemotherapy*, *47*(12), 3810–3814. https://doi.org/10.1128/AAC.47.12.3810-3814.2003

Lee, J., Han, J. J., Altwerger, G., & Kohn, E. C. (2011). Proteomics and biomarkers in clinical trials for drug development. *Journal of Proteomics*, *74*(12), 2632–2641. https://doi.org/10.1016/j.jprot.2011.04.023

Majoul, T., Bancel, E., Triboï, E., Ben Hamida, J., & Branlard, G. (2004). Proteomic analysis of the effect of heat stress on hexaploid wheat grain: Characterization of heat-responsive proteins from non-prolamins fraction. *PROTEOMICS*, *4*(2), 505–513. https://doi.org/10.1002/pmic.200300570

Monsarrat, B., Promé, J. C., Labarre, J. F., Sournies, F., & Van de Grampel, J. C. (1980). Mass spectrometry as a technique for testing the purity of drugs for biological use: The case of new antitumor cyclophosphazenes. *Biomedical Mass Spectrometry*, *7*(9), 405–409. https://doi.org/10.1002/bms.1200070910

Nakayasu, E. S., Sobreira, T. J. P., Torres, R., Ganiko, L., Oliveira, P. S. L., Marques, A. F., & Almeida, I. C. (2012). Improved proteomic approach for the discovery of potential vaccine targets in *Trypanosoma cruzi*. *Journal of Proteome Research*, *11*(1), 237–246. https://doi.org/10.1021/pr200806s

Nilsson, C. L., Larsson, T., Gustafsson, E., Karlsson, K. A., & Davidsson, P. (2000). Identification of protein vaccine candidates from *Helicobacter pylori* using a preparative two-dimensional electrophoretic procedure and mass spectrometry. *Analytical Chemistry*, *72*(9), 2148–2153. https://doi.org/10.1021/ac9912754

Paulo, J. A., Kadiyala, V., Banks, P. A., Steen, H., & Conwell, D. L. (2012). Mass spectrometry-based proteomics for translational research: A technical overview. *Yale Journal of Biology and Medicine*, *85*(1), 59–73.

Paweletz, C. P., Charboneau, L., Bichsel, V. E., Simone, N. L., Chen, T., Gillespie, J. W., Emmert-Buck, M. R., Roth, M. J., Petricoin III, E. F., & Liotta, L. A. (2001). Reverse phase protein microarrays which capture disease progression show activation of pro-survival pathways at the cancer invasion front. *Oncogene*, *20*(16), 1981–1989. https://doi.org/10.1038/sj.onc.1204265

Petricoin, E. F., & Liotta, L. A. (2003). Clinical applications of proteomics. *Journal of Nutrition*, *133*(7), 2476S–2484S. https://doi.org/10.1093/jn/133.7.2476S

Pisitkun, T., Johnstone, R., & Knepper, M. A. (2006). Discovery of urinary biomarkers. *Molecular & Cellular Proteomics: MCP*, *5*(10), 1760–1771. https://doi.org/10.1074/mcp.R600004-MCP200

Ray, S., Reddy, P. J., Choudhary, S., Raghu, D., & Srivastava, S. (2011). Emerging nanoproteomics approaches for disease biomarker detection: A current perspective. *Journal of Proteomics*, *74*(12), 2660–2681. https://doi.org/10.1016/j.jprot.2011.04.027

Ray, S., Reddy, P. J., Jain, R., Gollapalli, K., Moiyadi, A., & Srivastava, S. (2011). Proteomic technologies for the identification of disease biomarkers in serum: Advances and challenges ahead. *PROTEOMICS*, *11*(11), 2139–2161. https://doi.org/10.1002/pmic.201000460

Reddy, P. J., Jain, R., Paik, Y.-K., Downey, R., Ptolemy, A. S., Ozdemir, V., & Srivastava, S. (2011). Personalized medicine in the age of pharmacoproteomics: A close up on India and need for social science engagement for responsible innovation in post-proteomic biology. *Current Pharmacogenomics and Personalized Medicine*, *9*(1), 67–75. https://doi.org/10.2174/187569211794728850

Rodríguez-Ortega, M. J., Norais, N., Bensi, G., Liberatori, S., Capo, S., Mora, M., Scarselli, M., Doro, F., Ferrari, G., Garaguso, I., Maggi, T., Neumann, A., Covre, A., Telford, J. L., & Grandi, G. (2006). Characterization and identification of vaccine candidate proteins through analysis of the group A Streptococcus surface proteome. *Nature Biotechnology*, *24*(2), 191–197. https://doi.org/10.1038/nbt1179

Severin, A., Nickbarg, E., Wooters, J., Quazi, S. A., Matsuka, Y. V., Murphy, E., Moutsatsos, I. K., Zagursky, R. J., & Olmsted, S. B. (2007). Proteomic analysis and identification of *Streptococcus pyogenes* surface-associated proteins. *Journal of Bacteriology, 189*(5), 1514–1522. https://doi.org/10.1128/JB.01132-06

Sleno, L., & Emili, A. (2008). Proteomic methods for drug target discovery. *Current Opinion in Chemical Biology, 12*(1), 46–54. https://doi.org/10.1016/j.cbpa.2008.01.022

Traxler, P., Bold, G., Buchdunger, E., Caravatti, G., Furet, P., Manley, P., O'Reilly, T., Wood, J., & Zimmermann, J. (2001). Tyrosine kinase inhibitors: From rational design to clinical trials. *Medicinal Research Reviews, 21*(6), 499–512. https://doi.org/10.1002/med.1022

Walters, M. S., & Mobley, H. L. (2010). Bacterial proteomics and identification of potential vaccine targets. *Expert Review of Proteomics, 7*(2), 181–184. https://doi.org/10.1586/epr.10.12

Yang, M.-F., Liu, Y.-J., Liu, Y., Chen, H., Chen, F., & Shen, S.-H. (2009). Proteomic analysis of oil mobilization in seed germination and postgermination development of *Jatropha curcas*. *Journal of Proteome Research, 8*(3), 1441–1451. https://doi.org/10.1021/pr800799s

Zhang, A., Xie, C., Chen, H., & Jin, M. (2008). Identification of immunogenic cell wall-associated proteins of *Streptococcus suis* serotype 2. *Proteomics, 8*(17), 3506–3515. https://doi.org/10.1002/pmic.200800007

Zwick, E., Bange, J., & Ullrich, A. (2002). Receptor tyrosine kinases as targets for anticancer drugs. *Trends in Molecular Medicine, 8*(1), 17–23. https://doi.org/10.1016/S1471-4914(01)02217-1

Exercises 19.1

1. Vaccine targets should have which of the following properties?
   a. High antigenicity
   b. High immunogenicity
   c. Low persistence
   d. None of the above

2. Multi-protein drug targeting is better than single protein targeting because?
   a. Reduces toxic effects by the high drug dosage
   b. Increases the pathogen sensitivity to the drug
   c. Decreases the pathogen sensitivity to the drug
   d. Both a and b

3. HIV vaccine development is challenging because?
   a. Most of the HIV proteins share sequence similarity with human proteins
   b. HIV genes undergo spontaneous mutations which changes the antigens
   c. Vaccines cannot be developed against viruses
   d. Both a and b

4. Personalized medicine is gaining importance because of?
   a. Variations in the patient's genetic make-up and response to the drugs
   b. The identity/similarity of the same protein is less or negligible among the people
   c. Risk of disease incidence varies from person to person
   d. Both a and b

5. What is the molecular marker which is used for detecting the response of a patient on a drug treatment is called?
   a. Diagnostic marker
   b. Prognostic marker
   c. Predictive marker
   d. None of the above

6. Which among the following is *not* an application of protein microarrays?
   a. Studying protein interactions
   b. Identifying post-translational modifications
   c. Providing gene expression profiles
   d. Identifying novel protein-drug interactions

7. If a human proteome array contains several full-length proteins with GST tag, ideally with what antibody would you perform quality control checks?
   a. Anti-human IgG
   b. Anti-GST antibody
   c. Anti-flag antibody
   d. Cocktail of antibodies

8. State whether the following statement is true or false:
   Using biological networks, it is possible to find alternatives to drug targets by merely comparing their target networks.
   a. True
   b. False

9. You wish to study drug resistance in an organism via proteomics approach. Which of the following techniques will enable identification of proteomic alterations post-drug treatment?
   a. Whole cell lysis of the organism followed by SDS-PAGE
   b. Isolation of total DNA from the organism
   c. Two-dimensional gel electrophoresis of the organism with and without drug treatment
   d. Isoelectric focusing of the protein pellets from the organism with and without drug treatment

10. SILAC can be used to study which of the following?
    a. Protein-protein interaction
    b. Differential expression
    c. Signaling transduction pathways
    d. All of the above

11. Let's suppose, proteins X, Y, and Z have been identified as potential biomarkers for ovarian cancer in discovery phase proteomics analysis. Receiver operating characteristic (ROC) curve analysis has been performed to determine the specificity and sensitivity of the identified marker proteins. The AUC (area under the curve) for the proteins X, Y, and Z were observed to be 0.71, 0.93, and 0.65, respectively. Based on the AUC values which of the following ROC curves belong to which protein? Also, comment on the significance of these proteins in this diagnostic test.

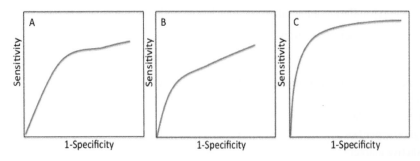

Answers

1. b
2. d
3. b
4. a
5. c
6. c
7. b
8. a
9. c
10. d
11. ROC Curve A represents the results obtained from the protein X.
    ROC Curve B represents the results obtained from the protein Z.
    ROC Curve C represents the results obtained from the protein Y.
    The most significant among these proteins is protein Y followed by X and Z.

20

Future of Proteomics for Clinical Applications

Preamble

Over the last decade, proteomics has been successful in integrating different areas of clinical research including biomarker discovery, disease pathogenesis, mechanism and targets of drug action, host-pathogen interactions and identification of potential drug/vaccine targets. Proteomics research has gained popularity because proteins are the eventual effector components of living organisms and have a direct correlation with disease pathobiology or physiological dysfunctions. To make the proteomics research more useful for real-life clinical applications, proteomic communities from different parts of the world have established Human Proteome Organization (HUPO) for driving the Human Proteome Project (HPP) and other related projects. HUPO aims to define the needs and guidelines for proteomic approaches in clinical research and preparation of global standards for sample processing, experimental procedures, data analysis, exchange and interpretation. In this chapter, the recent advancements in proteomics that are required to overcome the existing limitations will be discussed. Additionally, different proteomic projects and data repositories that serve as the major focal points for expanding proteomic research in clinical applications are also covered. A concise discussion on the status of proteomics research in India is also provided.

20.1 Recent Advancements in Proteomics

The 21st century is a witness of massive development in comprehensive proteomic analyses for potential clinical applications (Hanash, 2003; Petricoin & Liotta, 2003). Apart from the conventional gel and MS-based proteomic technologies, different advanced technological approaches have been introduced which may play some cardinal roles in clinical applications (Table 20.1).

Possible future applications of emerging proteomic technologies will be used for screening biological complexity, characterization of origin, pathobiology, disease biomarkers, and to develop preventive measures (drugs or vaccines) efficient for numerous classes of microorganisms. The introduction of new proteomic technologies with improved sensitivity and high-throughput capability will strengthen this dynamic field and circumvent existing limitations. It is equally important for the integration of proteomics discipline with other emerging research fields and collaborative research work at a global scale to translate the existing knowledge toward practical applications.

20.2 Worldwide Initiatives: Human Proteome Project and Other Related Projects

20.2.1 The Human Proteome Organization (HUPO) and Human Proteome Project (HPP)

The Human Proteome Organization (HUPO) is a global organization which aims to advance proteomics research at a global scale (www.hupo.org). In 2010, HUPO had initiated a global Human Proteome Project (HPP) for mapping all the proteins present in the human body. The major objective of HPP is to build the complete human protein map encoded by the genome, which may contribute to a better understanding of human diseases. To make these initiatives more comprehensive, different related initiatives,

DOI: 10.1201/9781003098645-25

TABLE 20.1

Definition and Description of Important Recent Proteomics Technologies

Technique	Description
Isobaric tag for relative and absolute quantitation (iTRAQ) [4- and 8-plex]	iTRAQ is a gel-free quantitative proteomics approach used for relative or absolute quantification of up to eight samples by using an isobaric tag consisting of a reporter group, a neutral balance portion, and a peptide reactive group. Unique mass values for the reporter ions allow quantification of different samples by MS/MS.
Isotope-coded affinity tags (ICAT) and visible ICAT (VICAT)	ICAT is a technique for quantitation of protein expression by differential labeling of two protein samples using "light" or "heavy" Cys-reactive alkylating reagents. VICAT is a new generation of ICAT reagent, which contains a visible tag that makes it possible to monitor electrophoretic movements of tagged peptides during separation. These quantitative proteomics approaches are very promising for the absolute quantification of specific target proteins in complex biological fluids.
Stable isotope labeling by amino acids in cell culture (SILAC) and super-SILAC	SILAC method depends on the metabolic incorporation of "light" and "heavy" forms of a particular amino acid (culture-derived isotope tags – CDIT), which provides stable and efficient labeling for protein quantification. Both relative and absolute quantitative proteome analysis is possible by CDITs. Super-SILAC is an advanced form of regular SILAC, which includes the application of a mixture of various SILAC-labeled cell lines as the internal standard for comprehensive comparative analysis.
Chip-based nano-LC-MS/MS	A recent advanced type of tandem MS (MS/MS) where the samples are separated in HPLC chips prior to MS/MS analysis. It is very effective method when a very minute number of samples are available and suitable for HT proteomic analysis. Specifically designed chips, which allow selective enrichment of phosphorylated/glycosylated peptides are very useful for PTM analysis.
Fourier transform ion cyclotron resonance MS) (FTICR-MS)	In FTICR-MS, the ions are trapped in a magnetic field combined with an electric field perpendicular to each other (penning trap), where they are excited to perform a cyclotron motion, and ion cyclotron resonance (ICR) signal is consequently converted to a frequency spectrum by Fourier transformation. FTICR-MS, although very complex in terms of data acquisition, has the highest resolving power and mass accuracy and, therefore very promising for clinical applications.
Multiple reaction monitoring (MRM) MS	MRM offers instantaneous quantification of several proteins in a fast sensitive and HT manner, which makes MS a striking substitute to antibody-based approaches for the detection of biomarker proteins. Instruments suitable for sequential analysis are required for conducting MRM-MS analysis. Information regarding the m/z of a particular precursor peptide and its selected product ions are required for performing MRM-MS.
Label-free MS	Label-free MS methods for quantification do not use labeling and stable isotopes to obtain quantitative information. Such quantitation methods provide accurate quantitative information and are very compatible for the determination of protein abundance changes in complex clinical samples.

like the International Cancer Genomic Consortium (ICGC), the Protein Data Bank (PDB) and the International Human Epigenomics Consortium (IHEC), are integrated with HPP (Legrain et al., 2011).

20.2.2 HUPO Plasma Proteome Project

In the year 2002, HUPO introduced the Human Plasma Proteome Project (HPPP) with multiple enduring ambitions. It included analysis of the entire protein component of human plasma and serum, identification of time-to-time biological/physiological variations in plasma proteome within individuals. It also covered the measurement of the magnitude of variations across individuals within same and different populations under normal and diseased conditions, which are very important for designing proper clinical studies (Omenn, 2007). As a further step of this initiative, the following phase of the HUPO Plasma Proteome Project (PPP-2) has been commenced in 2008 (17 August 2008, at the 7th World Congress of Proteomics in Amsterdam) (Omenn et al., 2009).

20.2.3 The PeptideAtlas Project

PeptideAtlas has been developed to annotate and validate eukaryotic genomes in a comprehensive manner by using protein identified in experimental studies. In April 2005, human build of PeptideAtlas has been constructed. It combined the peptide sequences of numerous proteins extracted from the different cell and tissue types identified in nearly 100 isolated proteomic experiments (Desiere et al., 2006). At present drafts for different human tissues/samples and animals is available at PeptideAtlas.

20.2.4 The HUPO Brain Proteome Project and Human Brain Proteome Atlas

The HUPO Brain Proteome Project (HUPO BPP) is an independent scheme under the support of HUPO that focuses on the analysis of the brain-related proteome project. The central ambition of BPP includes defining requirements and guidelines for proteomic approaches in brain research, in-depth investigation of neurodegenerative diseases and aging, identification of prognostic and diagnostic biomarkers, and supporting new diagnostic methods and medications (Hamacher et al., 2008). In the year 2010, HUPO BPP proposed the Human Brain Proteome Atlas (HBPA), and during the 10th HUPO Congress held in Geneva, in 2011, HBPA was officially launched with an intention to determine the proteomic contents of distinct parts of the human brain of healthy individuals and how it changes in response to aging and diseased conditions (Gröttrup et al., 2011).

20.2.5 Human Kidney and Urine Proteome Project (HKUPP)

In October 2005, Human Kidney and Urine Proteome Project (HKUPP) was launched for strengthening proteome level analysis of kidney functions and identification of reasons and mechanisms of chronic kidney diseases (CKDs). The ultimate goal of this project is to produce and exchange proteomics data on kidney and urine-related human diseases and the discovery of biomarkers and target molecules for new diagnostics and therapeutics (Yamamoto et al., 2008).

20.2.6 Human Liver Proteome Project (HLPP)

The scheme of the Human Liver Proteome Project (HLPP) was originally proposed in 2002 at the HUPO Workshop (Bethesda, MD, April 28–29, 2002) and as a consequence, a few months later the first HUPO Liver Proteome Workshop was organized (Beijing, October 22–24, 2002). The HLPP was the opening initiative of the HPP for the development of any organ-/tissue-specific proteome map with the purpose of making of ample protein diagram of the liver and the development of standard operating procedures for liver proteomic analysis. Additionally, establishment of an international liver tissue network, collection and distribution of liver samples from healthy subjects, and validation of new proteomic findings associated with human liver diseases were also included under the broad deliverables of this project (He, 2005).

20.2.7 HUPO Proteomics Standards Initiative (PSI)

In 2002, the HUPO Proteomics Standards Initiative (PSI) was introduced during the HUPO meeting in Washington (April 28–29, 2002) (http://www.psidev.info/), to establish the standards for the generation and representation of experimental data in proteomics, which can be followed by different proteomic research groups across the globe to provide uniformity in data comparison, exchange and interpretation, and to reduce technological variations (Orchard et al., 2003; Taylor et al., 2006).

20.3 Database Mining in Proteomics

Sharing of scientific data among different research groups across the world is essential for the improvement of overall research quality and progress and needs comprehensive database repositories and resources. A few such popular proteomic resources are mentioned here.

20.3.1 Uni-Prot KB/Swiss-Prot

The prime aspiration behind the development of Universal Protein Resource (UniProt) was the generation of a protein sequence repository containing inclusive coverage, which can be used as a centralized resource for protein annotation, integration and interpretation of data collected from several parts of the world. A combination of different groups from the European Bioinformatics Institute (EBI), the

Swiss Institute of Bioinformatics (SIB) and the Protein Information Resource (PIR) established UniProt Consortium, which played the prime role in the development of UniProt (Apweiler, 2004).

20.3.2 PRIDE

Proteomics identifications (PRIDE) database available at http://www.ebi.ac.uk/pride is a web-based query interface. Free web access or download and local installation are possible for the entire PRIDE database including the support tools and source codes, which makes it a very useful and popular resource for proteomics research (Martens et al., 2005). The major advantages of this online interface are that the data uploading procedure is very simple, and direct computational access is possible using a documented application programming interface provided in PRIDE. PRIDE is established as a useful resource for proteomics data for global accession.

20.3.3 Human Protein Atlas

Human Protein Atlas portal, released in August 2005, consists of 275 antibodies developed by the founders and over 400 other antibodies acquired from a variety of saleable antibody producers (Berglund et al., 2008). The antibody-based proteomics approaches make the researchers enable to carry out antibody-based functional studies of the entire proteome with a huge number of analyses to investigate the corresponding proteins. These initiatives are useful for the analysis of protein isoforms and comparative differential proteomic analysis of normal and diseased conditions with the objective to build a gene-centric human proteome map.

20.3.4 Tranche

Tranche is a redundant data repository completely dedicated to the storage, exchange and spreading of datasets among the proteomics communities. In order to expand its contents and incorporate different valuable tools for proteomics researchers, this data repository is strongly integrated with ProteomeCommons.org (Smith et al., 2011). The advantageous features of Tranche include its ability to handle very huge data files, licensing options, pre-publication access controls and authenticity.

20.4 Proteomics Research in India

In recent years, India has shown rapid progress in proteomics research along with the other omics-based disciplines (Sirdeshmukh, 2006, 2008) as apparent from the achievements in publications and patents. Quite a few research labs from India have started contributing significantly to global genomics and proteomics research and playing a critical role to advance this emerging field. Along with government support for basic and applied proteomics research, multiple national and international funding agencies are also coming forward to accelerate the pace of proteomics research in India. In recent years, different research groups from India have been identified as active contributors to proteomics and pharmacoproteomics with an intense application in personalized medicine and knowledge-based innovations (Reddy et al., 2011). A few years back, in order to enhance interactions among the proteomics communities across different parts of India and to improve the sharing of knowledge for the growth of this field, "Proteomics Society – India" was established (http://www.psindia.org/). India as well as other developing countries necessitate to be included as part of the developing international databanks to build some robust high-throughput platform for the acceleration of omics research. To this end, Data-Enabled Life Sciences Alliance (DELSA; www.delsall.org) is acting as a useful data repository for mining and exchange of biological data among different research communities at a global scale (Ozdemir et al., 2011).

Proteomics researchers from Indian Institute of Technology Bombay have developed a Virtual Proteomics laboratory (available at URL http://pe-iitb.vlabs.ac.in/) under the National Mission on Education through Information and Communication Technology (NME-ICT) scheme supported by the Ministry of Human Resource Development (MHRD), India (Ray, Koshy, Diwakar, et al., 2012; Ray, Koshy, Reddy, et al., 2012).

This is the first initiative from a developing country for the development of a web-based learning resource in the area of proteomics. This e-learning resource has been included as a component of the International Proteomics Tutorial Program conducted by HUPO and the European Proteomics Association (EuPA).

20.5 Conclusions

One of the most promising applications of proteomics in clinics could be the identification of clinically useful biomarkers and drug targets to improve diagnostics and therapeutics. High-throughput proteomic technologies capable of screening thousands of analytes within a very short period of time can further aid in the identification of new drug molecules. Despite having tremendous potential, the actual bed-side translation of the findings of the bench-side proteomic research is still limited. As a consequence, most of the proteomic applications in the clinical field are still at proof-of-concept levels. Therefore, the prime objective of modern proteomics research should be to drive the existing findings into realistic applications in diagnostics and therapeutics (Huber, 2003). If the basic limitations are circumvented successfully, it is anticipated that the proteomics platform will be very effective for clinical applications.

REFERENCES

Apweiler, R. (2004). UniProt: The Universal Protein knowledgebase. *Nucleic Acids Research*, *32*(90001), 115D–119D. https://doi.org/10.1093/nar/gkh131

Berglund, L., Björling, E., Oksvold, P., Fagerberg, L., Asplund, A., Szigyarto, C. A.-K., Persson, A., Ottosson, J., Wernérus, H., Nilsson, P., Lundberg, E., Sivertsson, A., Navani, S., Wester, K., Kampf, C., Hober, S., Pontén, F., & Uhlén, M. (2008). A genecentric Human Protein Atlas for expression profiles based on antibodies. *Molecular & Cellular Proteomics: MCP*, *7*(10), 2019–2027. https://doi.org/10.1074/mcp.R800013-MCP200

Desiere, F., Deutsch, E. W., King, N. L., Nesvizhskii, A. I., Mallick, P., Eng, J., Chen, S., Eddes, J., Loevenich, S. N., & Aebersold, R. (2006). The PeptideAtlas project. *Nucleic Acids Research*, *34*(Database issue), D655–D658. https://doi.org/10.1093/nar/gkj040

Gröttrup, B., Marcus, K., Grinberg, L. T., Lee, S. K., Meyer, H. E., & Park, Y. M. (2011). Creating a human brain proteome atlas—14th HUPO BPP workshop September 20–21, 2010, Sydney, Australia. *Proteomics*, *11*(16), 3269–3272. https://doi.org/10.1002/pmic.201190076

Hamacher, M., Marcus, K., Stephan, C., Klose, J., Park, Y. M., & Meyer, H. E. (2008). HUPO Brain Proteome Project: Toward a code of conduct. *Molecular & Cellular Proteomics: MCP*, *7*(2), 457.

Hanash, S. (2003). Disease proteomics. *Nature*, *422*(6928), 226–232. https://doi.org/10.1038/nature01514

He, F. (2005). Human liver proteome project: Plan, progress, and perspectives. *Molecular & Cellular Proteomics: MCP*, *4*(12), 1841–1848. https://doi.org/10.1074/mcp.R500013-MCP200

Huber, L. A. (2003). Is proteomics heading in the wrong direction? *Nature Reviews. Molecular Cell Biology*, *4*(1), 74–80. https://doi.org/10.1038/nrm1007

Legrain, P., Aebersold, R., Archakov, A., Bairoch, A., Bala, K., Beretta, L., Bergeron, J., Borchers, C. H., Corthals, G. L., Costello, C. E., Deutsch, E. W., Domon, B., Hancock, W., He, F., Hochstrasser, D., Marko-Varga, G., Salekdeh, G. H., Sechi, S., Snyder, M., … Omenn, G. S. (2011). The human proteome project: Current state and future direction. *Molecular & Cellular Proteomics: MCP*, *10*(7), M111.009993. https://doi.org/10.1074/mcp.M111.009993

Martens, L., Hermjakob, H., Jones, P., Adamski, M., Taylor, C., States, D., Gevaert, K., Vandekerckhove, J., & Apweiler, R. (2005). PRIDE: The proteomics identifications database. *Proteomics*, *5*(13), 3537–3545. https://doi.org/10.1002/pmic.200401303

Omenn, G. S. (2007). THE HUPO Human Plasma Proteome Project. *Proteomics. Clinical Applications*, *1*(8), 769–779. https://doi.org/10.1002/prca.200700369

Omenn, G. S., Aebersold, R., & Paik, Y.-K. (2009). 7(th) HUPO World Congress of Proteomics: Launching the second phase of the HUPOPlasma Proteome Project (PPP-2) 16–20 August 2008, Amsterdam, The Netherlands. *Proteomics*, *9*(1), 4–6. https://doi.org/10.1002/pmic.200800781

Orchard, S., Kersey, P., Hermjakob, H., & Apweiler, R. (2003). The HUPO Proteomics Standards Initiative Meeting: Towards common standards for exchanging proteomics data. *Comparative and Functional Genomics*, *4*(1), 16–19. https://doi.org/10.1002/cfg.232

Ozdemir, V., Rosenblatt, D. S., Warnich, L., Srivastava, S., Tadmouri, G. O., Aziz, R. K., Reddy, P. J., Manamperi, A., Dove, E. S., Joly, Y., Zawati, M. H., Hızel, C., Yazan, Y., John, L., Vaast, E., Ptolemy, A. S., Faraj, S. A., Kolker, E., & Cotton, R. G. H. (2011). Towards an ecology of collective innovation: Human Variome Project (HVP), Rare Disease Consortium for Autosomal Loci (RaDiCAL) and Data-Enabled Life Sciences Alliance (DELSA). *Current Pharmacogenomics and Personalized Medicine, 9*(4), 243–251. https://doi.org/10.2174/187569211798377153

Petricoin, E. F., & Liotta, L. A. (2003). Clinical applications of proteomics. *Journal of Nutrition, 133*(7), 2476S–2484S. https://doi.org/10.1093/jn/133.7.2476S

Ray, S., Koshy, N. R., Diwakar, S., Nair, B., & Srivastava, S. (2012). Sakshat Labs: India's virtual proteomics initiative. *PLoS Biology, 10*(7), e1001353. https://doi.org/10.1371/journal.pbio.1001353

Ray, S., Koshy, N. R., Reddy, P. J., & Srivastava, S. (2012). Virtual Labs in proteomics: New E-learning tools. *Journal of Proteomics, 75*(9), 2515–2525. https://doi.org/10.1016/j.jprot.2012.03.014

Reddy, P. J., Jain, R., Paik, Y.-K., Downey, R., Ptolemy, A. S., Ozdemir, V., & Srivastava, S. (2011). Personalized medicine in the age of pharmacoproteomics: A close up on India and need for social science engagement for responsible innovation in post-proteomic biology. *Current Pharmacogenomics and Personalized Medicine, 9*(1), 67–75. https://doi.org/10.2174/187569211794728850

Sirdeshmukh, R. (2006). Fostering proteomics in India. *Journal of Proteome Research, 5*(11), 2879. https://doi.org/10.1021/pr062768+

Sirdeshmukh, R. (2008). Proteomics in India: An overview. *Molecular & Cellular Proteomics: MCP, 7*(7), 1406–1407.

Smith, B. E., Hill, J. A., Gjukich, M. A., & Andrews, P. C. (2011). Tranche distributed repository and ProteomeCommons.org. *Methods in Molecular Biology (Clifton, N.J.), 696*, 123–145. https://doi.org/10.1007/978-1-60761-987-1_8

Taylor, C. F., Hermjakob, H., Julian, R. K., Garavelli, J. S., Aebersold, R., & Apweiler, R. (2006). The work of the Human Proteome Organisation's Proteomics Standards Initiative (HUPO PSI). *Omics: A Journal of Integrative Biology, 10*(2), 145–151. https://doi.org/10.1089/omi.2006.10.145

Yamamoto, T., Langham, R. G., Ronco, P., Knepper, M. A., & Thongboonkerd, V. (2008). Towards standard protocols and guidelines for urine proteomics: A report on the Human Kidney and Urine Proteome Project (HKUPP) symposium and workshop, 6 October 2007, Seoul, Korea and 1 November 2007, San Francisco, CA, USA. *Proteomics, 8*(11), 2156–2159. https://doi.org/10.1002/pmic.200800138

Exercises 20.1

1. Which of the following is the primary proteome project developed by HUPO?
   a. Kidney and Urine Proteome Project
   b. Plasma Proteome Project
   c. Human Proteome Project
   d. Liver Proteome Project

2. What was the first human proteome project for human organs/tissues?
   a. Liver Proteome Project
   b. Plasma Proteome Project
   c. Kidney and Urine Proteome Project
   d. None of the above

3. The HUPO Proteomics Standards Initiative (PSI) was established in which year?
   a. 2000
   b. 2003
   c. 2001
   d. 2002

4. UniProt Consortium consists of which of the following?
   a. European Bioinformatics Institute (EBI)
   b. Swiss Institute of Bioinformatics (SIB)
   c. Protein Information Resource (PIR)
   d. All of the above

5. When was the Human Protein Atlas portal released?
   a. July 2005
   b. August 2005
   c. August 2006
   d. April 2007

6. Which of the following will be essential to expand the application of clinical proteomics?
   a. Development of more HT and ultra-sensitive detection techniques
   b. Collaborative proteomics initiatives at a global level
   c. Integration of proteomics research with other emerging research fields
   d. All of the above

7. The Data-Enabled Life Science Alliance (DELSA) encompasses which of the following aspect(s)?
   a. Sharing scientific findings
   b. Collaborative approaches
   c. Data computation and cloud-based storage
   d. All of the above

8. MRM is a mass spectrometric-based assay that enables which of the following?
   a. High-throughput data integration and analysis
   b. Pathway analysis of multiple reactions
   c. LC-MS based quantification of peptides
   d. All of the above

9. Advantage(s) of label-free detection techniques include which of the following?
   a. Enables the monitoring of the sample with minimalistic modification
   b. Doesn't use kinetic assays and can give molecular structural information
   c. Uses light sensitivity and energy transfer
   d. All of the above

10. The Human Proteome Project launched in 2010 by the Human Proteome Organization (HUPO) was initiated to
    a. Annotate the enzymes coded by the Human
    b. To generate map of the protein based molecular architecture by characterizing all 25,000 known human gene of the genome
    c. To map the proteins coded by the human chromosomes
    d. All of the above

11. In September 2010, the HUPO Initiatives Committee at the HUPO 9th Annual World Congress approved a new initiative which enables to standardize efforts and to make optimum use of the proteomics data acquired in model organisms. Do some literature search and find out which HUPO initiative are we talking about here. Why is there such a need for studying model organism proteome and what are the objectives of this initiative?

Answers

1. c
2. a
3. d
4. d
5. b
6. d
7. d
8. c
9. a
10. d
11. The initiative, which has been described above, is called initiative in model organism proteomes (iMOP).

    The model organisms serve as perfect platform for studying certain disease pathways because of their easiness to grow in the labs, short generation time, and above all the conservation of disease pathways between model organisms and human beings. These kinds of studies can also be used for the development of drugs.

    The main objectives of iMOP are as follows:

    1. Integration of various model organism research teams into a model organism proteomics community.
    2. Adoption of HUPO standards and practices.
    3. Integration and linking of proteome and organism-specific databases.
    4. Development of software tools in order to navigate these databases.

21

Challenges in Clinical Proteomics

Preamble

Successful completion of the human genome sequence project catalyzed the progress of proteomics research in different disciplines of modern science. Starting from the mid-1990s, in due course, this promising field has propelled its expansion in nearly every aspect of life science research. The emerging proteomics techniques have incredible potential to offer a plethora of new information to accelerate the pace of basic and applied clinical research. However, there are quite a few basic limitations of proteome level research, mostly due to the fragile nature of proteins leading to substantial losses during sample collection and processing steps, post-translational modifications, presence of multiple isoforms of same proteins, and complexity and wide dynamic range of protein concentrations. Moreover, there is no direct amplification strategy in proteome-level research equivalent to PCR, which is used for gene amplification. In this chapter, the challenges associated with the commonly used proteomics techniques will be discussed in light of clinical applications.

Terminology

- **Proteomics:** Proteomics is a comprehensive study of the expression, localization, interaction, and post-translational modifications of the whole set of proteins encoded by the genome in a given condition.
- **Clinical proteomics:** The most promising sub-discipline of proteomics, which deals with different possible clinical applications of proteomics, including disease biomarker discovery, the study of disease pathogenesis, drug action, host-pathogen interactions, and identification of potential drug/vaccine targets.
- **Gel-based proteomics:** Protein separation techniques where polyacrylamide gels are applied for the separation of proteins present in complex mixtures. SDS-PAGE, native-PAGE, 2-DE, and DIGE are the widely adopted gel-based proteomics techniques.
- **MS-based proteomics:** Combination of the most useful analytical technologies for accurate mass measurement. Although there are many MS-based tools, the most commonly used MS platforms are ESI-Q-TOF-MS/MS, MALDI-TOF and MALDI-TOF-TOF. More hybrid tandem MS technologies have been introduced in recent years.
- **Protein microarrays:** High-throughput proteomics techniques where the large number of proteins are concurrently immobilized on a glass or polyacrylamide gel pad surfaces, while the target proteins dissolved in solution are allowed to pass through to investigate molecular interactions.
- **Label-free detection techniques:** Label-free methods for quantification eliminate the need for labeling the query molecules to obtain quantitative information. Such quantitation approaches provide accurate information for the protein abundance changes in complex samples.
- **Nanoproteomics:** Integration of different nanotechnological approaches in proteomics generated this amalgamated analytical platform which effectively improved the limit of detection, dynamic range, detection speed and multiplexing power of different conventional proteomics techniques.
- **Biomarkers:** Biomarkers are indicator biomolecules that help to detect diseased states at an early stage, make discrimination between different diseases, and provide useful information for monitoring the progression/severity of the disease.

DOI: 10.1201/9781003098645-26

- **Membrane proteins:** Proteins present in biological membranes are called membrane proteins, which are very difficult to isolate and study because of their hydrophobic properties and comparatively low concentration.
- **Post-translational modifications (PTMs):** PTMs are alterations in the polypeptide chain generated by either the addition or removal of diverse chemical moieties, proteolytic cleavage, or covalent cross-links between different domains of the protein, which can effectively change the molecular functions of the proteins. Most commonly occurring PTMs include phosphorylation, O-glycosylation, sulfation, nitration and acylation.

21.1 Technical Challenges Associated with Different Proteomic Technologies

There are some inherent challenges in the field of proteomics due to the following reasons:

1. Fragile nature of proteins.
2. Lack of any protein amplification method (like PCR used for gene amplification).
3. Difficulties in extraction and isolation of proteins from specific organelles.
4. Presences of various isoforms of a single protein.
5. Massive variation in proteome with time within the same individual.
6. Variations among the individuals of the same or different populations.
7. The wide dynamic range of protein concentrations in biological fluids.

There are several technological limitations associated with proteomics approaches as summarized in Table 21.1.

TABLE 21.1

Limitations Associated with Different Proteomics Technologies

Type	Commonly Used Techniques	Limitations	Recent Advancements
(a) Gel-based proteomics	1D SDS-PAGE, native-PAGE, 2-DE	• Poor reproducibility • Inadequate sensitivity and dynamic range (10^3–10^4) • Insufficient coverage of complex proteome • Low-throughput • Biasness in analysis process • Lengthy experimental process • Reliance on performer's technical skill	2D-DIGE: Better reproducibility and sensitivity (Unlü et al., 1997) Post-electrophoresis epicoccone fluorescent dyes like lightning fast and deep purple: Increased sensitivity, dynamic range, and coverage (Miller et al., 2006) • Activity-based protein profiling (enzyme-targeting probes) information regarding protein activity (Hu et al., 2004)
(b) MS-based proteomics	MALDI-TOF-TOF, ESI-Q-TOF, ESI-TRAP, MALDI-QUAD-TOF	• Narrow dynamic ranges (10^2–10^4) • Inadequate coverage • Low-throughput • Overfitting of the data • Machine fluctuation • Instrument noise and contaminants in spectrum • Dearth of standard procedure for analysis and interpretation of MS and MS/MS spectrum (deVera et al., 2006; Patterson, 2003)	• CDIT and Super-SILAC: large-scale quantitative proteomics • TMT and iTRAQ (4- and 8-plex): multiplexing • Label-free LC-MS/MS: superior quantitative accuracy • Chip-based and Nano-LC-MS: low sample consumption • FTICR, LTQ-FT; improved sensitivity • MRM MS: large-scale biomarker discovery (Qian et al., 2006)

(Continued)

TABLE 21.1 (*Continued*)

Limitations Associated with Different Proteomics Technologies

Type	Commonly Used Techniques	Limitations	Recent Advancements
(c) Array-based proteomics	Protein microarrays, antibody microarrays, reverse-phase microarrays	• Protein array designing difficulties • Problems in acquisition, arraying, and stable attachment of proteins to array surfaces • Inadequate sensitivity to detect very weak interactions and low-abundance analytes • Miniaturization of assays and protein dehydration • Non-specific binding • Unavailability of highly specific antibodies • Lack of direct correlation between protein abundance and activity (Kodadek, 2001; Talapatra et al., 2002)	• On-chip synthesis of protein; protein arraying by cell-free expression • Combination with label-free detection techniques: real-time sensitive detection • Use of nanoparticles for signal amplification: better sensitivity
(d) Label-free proteomics	SPR, SPRi, ellipsometry- and interference-based techniques, microcantilevers	• Reduced sensitivity and specificity when complex samples are analyzed (Ray, Mehta, et al., 2010; Yu et al., 2006) • Costly fabrication techniques • Morphological anomalies of sample spots • Insufficient knowledge regarding the precise working mechanism	New label-free methods; SPR-MS, backscattering interferometry, Brewster's angle straddle interferometer, UV fluorimetry, tagged-internal standard assay, and spectral-domain optical coherence phase microscopy: improvement of sensitivity and HT capability (Ray, Mehta, et al., 2010)
(e) Nanoproteomics	Carbon nanotubes, nanowires, silicon nanowire field effect transistor, quantum dots, gold nanoparticles, and nanocages	• Toxicity, biosafety, and biocompatibility issues • Inadequate knowledge on the precise mechanism of action • Insolubility in biologically compatible buffers • Short lifetime • Presence of metallic impurities • Lack of standard protocol for determining degree of purity of synthesized nanotubes and nanowires (Ray, Chandra, et al., 2010; Ray, Reddy, Choudhary, et al., 2011)	• Encapsulating shell and capping materials: longer lifetime • Combinations with immunoassays • New cost-effective fabrication techniques (Ray, Reddy, Choudhary, et al., 2011)

21.2 Challenges in Biomarker Discovery: Detection of Low-Abundant Proteins

Biomarkers are indicator biomolecules that help to detect diseased states at an early stage, discriminate between different diseases and useful for monitoring the progression/severity of a disease. In spite of diverse advancements, there are several biological and technological limitations in the present proteomics technologies that are regularly applied for the discovery of disease-related marker proteins. Pre-analytical variations introduced during sample collection, handling and storage process are also detrimental for the screening of true biomarkers. Additionally, the complexity of biological sample, very dynamic range of protein concentrations, presence of high-abundance proteins masking low-abundance marker proteins, high levels of salts, and other interfering compounds in most of the biological specimens, insufficient sensitivity of the detection technology, and paucity of throughput and multiplexed detection ability are the major obstacles for the direct application of proteomics technologies in clinics (Ray, Reddy, Jain, et al., 2011).

With time, different combinations of separation, detection, and labeling strategies such as SCX (separation), ICAT, iTRAQ, TMT (labeling), nanoparticles like nanowires, nanotubes, quantum dots (signal amplification and enrichment of low-abundance proteins), surmount the basic technological limitations associated with existing proteomics approaches. Selective enrichment of low-abundance biomarkers and protection of degradable proteins are fascinated by the application of core-shell hydrogel particles functionalized with various affinity selector baits and size exclusion exteriors. When the hydrogel particles are kept in biological fluids, those particles selectively entrap low-abundance biomarkers in their baits and increase the effective concentration of those analytes (Longo et al., 2009) (Figure 21.1).

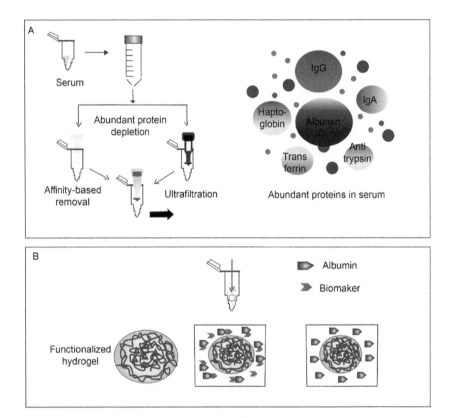

FIGURE 21.1   Removal of high-abundant proteins. (A) Serum sample preparation for proteomic analysis. The inset shows 6 high-abundant proteins, namely albumin, IgG, IgA, haptoglobin, transferrin, and anti-trypsin, which make it very challenging to detect the relatively low-abundant protein biomarkers. (B) Application of hydrogel particles for selective enrichment and preservation of low-abundant biomarkers.

21.3 Challenges in Membrane Proteomics

Investigation of the proteins present in biological membranes is very informative but challenging due to the hydrophobic properties of the membrane proteins. Additionally, most of the membrane proteins are low-abundant in nature. For proteome-level analysis, membrane proteins must be solubilized from the lipid layers, but usually, the solubility of membrane proteins is low at their isoelectric points. Due to the presence of multiple hydrophobic domains, membrane proteins tend to aggregate and subsequently precipitate out during commonly used proteomic sample preparation approaches (Helbig et al., 2010).

For enrichment of membrane proteins prior to the proteomic analyses, different fractionation methods, free-flow electrophoresis and two-phase partitioning of the membrane vesicles are employed. Additionally, different commercially available surfactants are being used for improving membrane protein solubilization to increase the coverage of total membrane protein identification.

21.4 Challenges in Analysis of PTMs

The biological activities and molecular function of the majority of the eukaryotic proteins depend on the post-translational modifications (PTMs). Efficient and sensitive methods for a large-scale study of PTMs are still lacking (Mann & Jensen, 2003). Most commonly occurring PTMs include phosphorylation (+80 Da), O-glycosylation (>203 Da), sulfation (+80 Da), nitration (+45 Da), and acylation (>200 Da). The analysis of PTMs in proteomics is a challenging task since most of the PTMs are of low abundance and/or substoichiometric. MS-based analysis of PTM is very promising, however, the chemical stability of PTM is crucial for its efficient detection since many of the PTMs are very labile during MS and MS/MS analysis. Additionally, many PTMs are hydrophilic in nature, which makes PTM sample handling and purification extremely difficult prior to MS (Larsen et al., 2006).

21.5 Conclusions

The findings obtained from proteome-level research in the last decade have contributed significantly in unraveling the various unexplored secrets of human diseases and paved the way of proteomics in different applications of clinical research including elucidation of the mechanism of drug action, identification of drug and vaccine targets, the establishment of diagnostic and prognostic biomarkers. However, different technological limitations associated with the most common candidates available in the proteomic toolbox are hindering the bed-side translation of proteomic approaches in real life. The impending future of this highly promising research field will depend on successful solution of the existing limitations and collaborative initiatives at the global level to prepare standard protocols for clinical proteomics researchers to avoid pre-and post-analytical variations.

REFERENCES

deVera, I. E., Katz, J. E., & Agus, D. B. (2006). Clinical proteomics: The promises and challenges of mass spectrometry-based biomarker discovery. *Clinical Advances in Hematology & Oncology: H&O, 4*(7), 541–549.

Helbig, A. O., Heck, A. J. R., & Slijper, M. (2010). Exploring the membrane proteome—Challenges and analytical strategies. *Journal of Proteomics, 73*(5), 868–878. https://doi.org/10.1016/j.jprot.2010.01.005

Hu, Y., Huang, X., Chen, G. Y. J., & Yao, S. Q. (2004). Recent advances in gel-based proteome profiling techniques. *Molecular Biotechnology, 28*(1), 63–76. https://doi.org/10.1385/MB:28:1:63

Kodadek, T. (2001). Protein microarrays: Prospects and problems. *Chemistry & Biology, 8*(2), 105–115. https://doi.org/10.1016/s1074-5521(00)90067-x

Larsen, M. R., Trelle, M. B., Thingholm, T. E., & Jensen, O. N. (2006). Analysis of posttranslational modifications of proteins by tandem mass spectrometry. *BioTechniques, 40*(6), 790–798. https://doi.org/10.2144/000112201

Longo, C., Patanarut, A., George, T., Bishop, B., Zhou, W., Fredolini, C., Ross, M. M., Espina, V., Pellacani, G., Petricoin, E. F., Liotta, L. A., & Luchini, A. (2009). Core-shell hydrogel particles harvest, concentrate and preserve labile low abundance biomarkers. *PloS One, 4*(3), e4763. https://doi.org/10.1371/journal.pone.0004763

Mann, M., & Jensen, O. N. (2003). Proteomic analysis of post-translational modifications. *Nature Biotechnology, 21*(3), 255–261. https://doi.org/10.1038/nbt0303-255

Miller, I., Crawford, J., & Gianazza, E. (2006). Protein stains for proteomic applications: Which, when, why? *Proteomics, 6*(20), 5385–5408. https://doi.org/10.1002/pmic.200600323

Patterson, S. D. (2003). Data analysis—The Achilles heel of proteomics. *Nature Biotechnology, 21*(3), 221–222. https://doi.org/10.1038/nbt0303-221

Qian, W.-J., Jacobs, J. M., Liu, T., Camp, D. G., & Smith, R. D. (2006). Advances and challenges in liquid chromatography-mass spectrometry-based proteomics profiling for clinical applications. *Molecular & Cellular Proteomics: MCP*, 5(10), 1727–1744. https://doi.org/10.1074/mcp.M600162-MCP200

Ray, S., Chandra, H., & Srivastava, S. (2010). Nanotechniques in proteomics: Current status, promises and challenges. *Biosensors & Bioelectronics*, 25(11), 2389–2401. https://doi.org/10.1016/j.bios.2010.04.010

Ray, S., Mehta, G., & Srivastava, S. (2010). Label-free detection techniques for protein microarrays: Prospects, merits and challenges. *Proteomics*, 10(4), 731–748. https://doi.org/10.1002/pmic.200900458

Ray, S., Reddy, P. J., Choudhary, S., Raghu, D., & Srivastava, S. (2011). Emerging nanoproteomics approaches for disease biomarker detection: A current perspective. *Journal of Proteomics*, 74(12), 2660–2681. https://doi.org/10.1016/j.jprot.2011.04.027

Ray, S., Reddy, P. J., Jain, R., Gollapalli, K., Moiyadi, A., & Srivastava, S. (2011). Proteomic technologies for the identification of disease biomarkers in serum: Advances and challenges ahead. *PROTEOMICS*, 11(11), 2139–2161. https://doi.org/10.1002/pmic.201000460

Talapatra, A., Rouse, R., & Hardiman, G. (2002). Protein microarrays: Challenges and promises. *Pharmacogenomics*, 3(4), 527–536. https://doi.org/10.1517/14622416.3.4.527

Unlü, M., Morgan, M. E., & Minden, J. S. (1997). Difference gel electrophoresis: A single gel method for detecting changes in protein extracts. *Electrophoresis*, 18(11), 2071–2077. https://doi.org/10.1002/elps.1150181133

Yu, X., Xu, D., & Cheng, Q. (2006). Label-free detection methods for protein microarrays. *Proteomics*, 6(20), 5493–5503. https://doi.org/10.1002/pmic.200600216

Exercises 21.1

1. Which of the following is the major limitation of classical 2-DE?
   a. Dynamic range
   b. Sensitivity
   c. Gel-to-gel variations
   d. Low throughput

2. The completion of Human Genome Project in 21st century has seen the rise of numerous disciplines holistically studying various biomolecules that make up the living cells. These disciplines are suffixed – omics on the account of them presenting collective technologies to study the entire compendium of that given biomolecules at any given time. Match each of these "omics" discipline to the biomolecules that they majorly focus on.

Discipline	Biomolecules
(A) Genome	(i) mRNA
(B) Transcriptome	(ii) Metabolite
(C) Translatome	(iii) Protein
(D) Proteome	(iv) Phenotype
(E) Metabolome	(v) Ribosome
(F) Phenome	(vi) DNA

   a. A-v, B-i, C-vi, D-iii, E-ii, F-iv
   b. A-iv, B-ii, C-v, D-iii, E-i, F-vi
   c. A-vi, B-i, C-v, D-ii, E-iii, F-iv
   d. A-vi, B-i, C-v, D-iii, E-ii, F-iv

3. Which the followings is an existing limitation(s) of array-based proteomics?
   a. Unavailability of highly specific antibodies against all the target proteins
   b. Lack of direct correlation between protein abundance and activity
   c. Difficulties in stable attachment of proteins to array surfaces
   d. All of the above

4. Systems biology integrates several facets of biology, omics, and bioinformatics techniques. Which of the following approach(s) can be used in systems biology?

   a. Elucidation of transcription factors
   b. Network building using bioinformatics tools
   c. Metabolite profiling and metabolomics
   d. All of the above

5. Why is pre-fractionation of biological samples essential prior to the gel- or MS-based proteomic analysis?

   a. Low dynamic range of the proteomic technologies
   b. Complexity of the biological samples
   c. Both of the above
   d. None of the above

6. Why is the study of membrane proteomics challenging?

   a. Hydrophobic properties of membrane proteins
   b. Relatively low abundance of membrane proteins
   c. Poor solubility of membrane proteins under the chosen experimental conditions
   d. All of the above

7. Which of the following is a challenge(s) posed in protein microarrays?

   a. Diversity in protein sizes
   b. Spatial location of proteins
   c. Post-translational modifications
   d. Hydrophobic nature of some proteins
   e. All of the above

8. Complex proteomics studies unravelling the physiological pathways are very crucial if we are mapping protein-protein interaction networks. Which of the following methods is *not* used for protein-protein interaction studies?

   a. Surface plasmon resonance (SPR)
   b. Bio-layer interferometry (BLI)
   c. Protein microarray
   d. SDS-PAGE

9. Which of the following strategy(s) is often used during protein extraction from serum samples?

   a. Sonication
   b. Removal of high abundant proteins
   c. Desalting
   d. All of the above

10. Which of the following study(s) involves small molecule application of mass spectrometry?

   a. Drug development
   b. Forensics science
   c. Food safety analysis
   d. All of the above

11. While performing serum proteomic analysis to identify low-abundant potential biomarkers, high-abundant proteins mask the presence of the low-abundant proteins which leads to an insensitive assay. Below is shown a method which helps in enriching such interesting low-abundant proteins. Your task is to identify the method and basic components of the procedure while explaining the core principle involved in this technique.

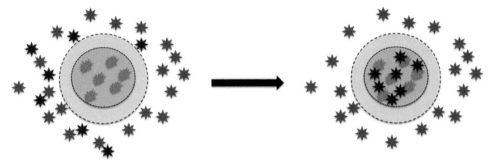

Answers

1. c
2. d
3. d
4. d
5. c
6. d
7. e
8. d
9. d
10. d
11. The method shown above depicts the enrichment of the low-abundant proteins in the serum samples by using core-shell hydrogel particle functionalized with affinity selector baits and size exclusion exteriors. The low-abundant proteins are selectively trapped when the serum is applied onto these components.

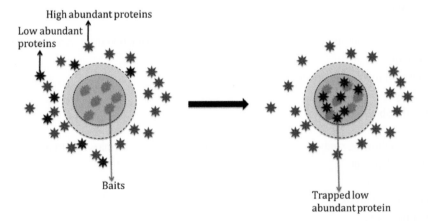

Index

A

α-helices, 6, 7
Aminosilane, 167, 180, 181, 186
Amplification, 149, 217, 232, 251, 252, 253, 254
Antigen-antibody reaction, 181, 216, 145, 148
Anti-GST antibody, 167, 179, 181, 187, 188, 240
Anti-human IgG, 183, 240
Anti-tag, 167
APTES, 167, 174, 177, 181
Artifacts, 38, 39, 55, 57, 59, 111, 223
Autoantibodies, 158, 159, 182, 183
Autoimmune, 154, 216, 231

B

β-sheets, 6, 7, 8
Bait protein, 135, 137, 139
Balance group, 113
Binding kinetics, 197, 206, 207, 211, 217, 222
Bio-barcode, 153, 156
Bioinformatics, 13, 14, 16, 17, 18, 135, 245, 245, 249, 257
Biological variation analysis, 45, 49, 65
b-ions, 101, 103
Biotin-avidin, 181, 186

C

CCD camera, 189, 193, 216, 218, 221, 222
Cell-free production, 161
Chirality, 4
Chromogenic, 190
Codon, 3, 70, 106
Co-factors, 162
Collision-induced dissociation, 100
Coomassie brilliant blue, 35, 63
Cross-linker, 166, 167, 179, 180, 181
Cross-reactivity, 148, 149, 156, 216
C-terminal GST, 181
Cyanine dyes, 16, 37, 45, 46, 47, 48, 63
Cytokine, 153
Cytotoxicity, 151

D

Databases, 17, 18, 103, 118, 126, 250
Daughter ions, 94, 95, 101
De novo sequencing, 17, 99, 106
Depletion columns, 87
Derivatization, 147
Desorption, 15, 79, 82, 89, 192, 205, 216, 221
Diagnostics, 217, 231, 232, 237, 245, 247

Difference gel electrophoresis, 13, 16, 25, 38
Differential expression analysis, 69
Differential in-gel analysis, 45, 65
Discriminant analysis, 70
Dissociation constant, 221, 228
Drug development, 232, 234, 235, 257
Dye-doped, 152, 156

E

Edman degradation technique, 18
Electrochemical, 192, 208, 225
ELISA, 111, 148
Ellipsometry, 192, 193, 194, 197, 200, 201, 203, 211, 253
Endosperm, 163
Expression vector, 146
Extended data analysis, 65, 68

F

Fabrication, 125, 199
Fluorescence, 37, 39, 50, 147, 150, 151, 152, 156, 190, 197, 208, 211, 225
Fluorescence-based, 37, 39, 150, 156, 157, 190
Fluorescent-tagged, 147
Fourier transformer ion cyclotron resonance (FT-ICR), 84, 86, 96, 244

G

Gateway cloning, 178
Genomics, 21, 79, 177, 236, 246
Glutathione-S-transferase (GST), 166
Gold-coated surfaces, 192, 196, 206, 215

H

HaloLink, 161, 170, 172, 174
Hexa-histidine, 165, 172
His6, 145, 147
Horseradish peroxidase, 183
Human Genome Project, 16, 17
Hydrophobic collapse, 8

I

ImageQuant TL software, 27
Immunoassay, 206, 209, 211, 253
Immunogenicity, 171, 239
Immunoprecipitation, 135, 138, 145, 221
In vitro translation, 163, 165, 168

Printed and bound by CPI Group (UK) Ltd, Croydon, CR0 4YY

24/10/2024

01778288-0008